Intelligence at the Edge
Using SAS® with the Internet of Things

Edited by
Michael Harvey

sas.com/books

The correct bibliographic citation for this manual is as follows: Harvey, Michael. 2020. *Intelligence at the Edge: Using SAS® with the Internet of Things*. Cary, NC: SAS Institute Inc.

Intelligence at the Edge: Using SAS® with the Internet of Things

Copyright © 2020, SAS Institute Inc., Cary, NC, USA

ISBN 978-1-64295-780-8 (Hardcover)
ISBN 978-1-64295-776-1 (Papberback)
ISBN 978-1-64295-777-8 (PDF)
ISBN 978-1-64295-778-5 (epub)
ISBN 978-1-64295-779-2 (kindle)

All Rights Reserved. Produced in the United States of America.

SAS Institute Inc., SAS Campus Drive, Cary, NC 27513-2414

February 2020

Contents

Preface

About the Internet of Things

Today we can hardly imagine a world without the internet. Whether we access information about a restaurant or product through our smart phone, manage our banking transactions through those same phones or our home computers, or plan a vacation through easy-to-use websites, we don't give a second thought to using it. The internet is always there, at our fingertips.

It was only sixty years ago that the preliminary research on packet switching began. Packet switching undergirds the TCP/IP protocols used by the internet. TCP/IP was designed to provide a more robust alternative network architecture than existing point-to-point computer networks. By 1971, fifteen sites were internetworking on the ARPANET, which became one of the first backbones of the internet. In 1989, Tim Berners-Lee, a computer scientist at CERN, invented the World Wide Web. He engineered a web browser to navigate this web which he released to the public in 1991. Twenty years later, 2.2 billion users (almost a third of the world's population) were using browsers to traverse the internet. They used these browsers on laptops, tablets, and smart phones (Miniwatts Marketing Group, 2019).

As you read these pages, we are experiencing a seismic shift in the way that we interact with the internet. It is no longer confined to using a web browser – it now touches almost every aspect of our day-to-day activity. The Internet of Things (IoT) takes the internetworked computer systems with which we are so familiar and attaches a plethora of devices, sensors, and objects in our world. Data is collected and processed in real time from these "things."

Those "things" are ubiquitous. Consider the following. Not long ago, my wife and I were driving back from a weekend getaway. She picked up her phone and asked her digital assistant, "How long will it take to arrive home?" Within seconds, the assistant gave a precise reply in hours and minutes. When we arrived home, the assistant's reply was off by a mere five minutes. (We encountered some congestion during the final mile of our trip.)

Think about that. While speeding down a four-lane highway, my wife used her phone, a "thing," to use the internet to ask an artificially intelligent digital assistant when we would arrive home. In a manner of seconds, that assistant used the phone's GPS to determine our current location and our current speed. Using our destination location, it calculated the distance that we wanted to travel. Accounting for current traffic conditions, it calculated the time it would take to arrive at our destination. In seconds. And the calculation was accurate within just a few minutes.

When was the Internet of Things born? Some say it was in the 1980s, when programmers at Carnegie Melon University connected a Coke machine to the internet to check whether a drink was available and was cold (Foote, 2016). Nevertheless, it was not until 1999 that the phrase "Internet of Things" was coined by Kevin Ashton, the Executive Director of Auto-ID

Labs at MIT. He proposed using Radio Frequency Identification (RFID) technology to tag devices so that computers could manage, track, and inventory them through the internet. Since then, things have been tagged with digital watermarking, bar codes, and QR codes, among other means (Foote, op cit). The number of things connected to the internet has grown at a breathtaking rate.

These days, Internet of Things involves not just collecting and tracking data, but also applying analytics to the data *as it is gathered* to intelligently manage those things. You can run analytics on systems that are physically close to the device collecting the data, that is, at the "edge," to enable faster decision making. SAS software enables organizations to collect and apply intelligence at the edge to better extract business value. Businesses translate that value into more efficient operations, lower costs, and a wider range of revenue streams. Thus, now when you collect data from sensors on trucks, trains, or jets, you can better predict when parts might fail. You then can use the results of your analysis to proactively stock the right parts and keep the vehicles operational, saving thousands of hours of downtime. You can monitor streaming data from factory equipment to more intelligently schedule maintenance. You can analyze streaming data from call center systems, news sites, and social media forums, and then integrate your analysis with issue detection processes to get a jump on corrective actions. Applying intelligence at the edge has become one of the highest priorities for businesses.

Foote, Keith D. 2016. "A Brief History of the Internet of Things," https://www.dataversity.net/brief-history-internet-things/.

Miniwatts Marketing Group, 2019. Internet Growth Statistics https://www.internetworldstats.com/emarketing.htm.

About This Book

This book is written for a general audience who wants to learn more about this rapidly changing field. Some technical knowledge of computing and statistics is useful to fully grasp the terms and concepts covered, but it is not essential to comprehend the information provided. The book describes how SAS applies analytics to derive business value from the Internet of Things.

At the heart of that endeavor is SAS Event Stream Processing. The first chapter explains how that product works, provides some simple examples, and briefly covers a scenario where an ESP server analyzes edge data and sends results to another ESP server running at the data center. It explains the distinction between static data, which sits unchanging in a database or repository, and streaming data, which continuously flows from the world into software applications.

SAS Event Stream Processing gets data from the world primarily through connectors and adapters, which are covered in Chapter 2. It explains how SAS Event Stream Processing uses connectors and adapters to communicate with the message fabrics, databases, and other protocols used to handle data transmitted from the "things" in the IoT. It includes an example that shows how connectors work.

Chapter 3 describes how to apply analytics to streaming data. It describes the IoT analytics life cycle emphasizing the analytics. It covers the algorithms that you can apply to streaming data and provides brief examples of two of those algorithms applied to data from the edge.

How can you effectively deploy and manage ESP servers at the edge? The answer is by using SAS Event Stream Manager. Chapter 4 describes how to monitor and manage your SAS Event Stream Processing environment through this web-based client. It shows how SAS Event Stream Manager works with SAS Model Manager to enable you to use the most current champion model in your environment. It also provides a step-by-step example to create a SAS Event Stream Manager deployment, associate an ESP server with it, and run, monitor, and stop the associated job.

What is an IoT reference architecture, and how does SAS Event Stream Processing fit into it? Chapter 5 covers that and the deployment considerations for the edge and cloud. It discusses the IoT life cycle emphasizing the life cycle. It provides a detailed example of using SAS Event Stream Processing with other SAS products within an IoT reference architecture.

The final chapters cover use cases of analyzing data from the Internet of Things with SAS Event Stream Processing. Chapter 6 gives an overview of applying AI to streaming data from the IoT using SAS Event Stream Processing. It also provides a detailed example of using SAS Visual Data Mining and Machine Learning to build a model and then deploying that model in SAS Event Stream Processing to score streaming data. Chapter 7 introduces the topic of geofencing and explains how the Geofence window of SAS Event Stream Processing enables it in real-time. Chapter 8 tells how SAS Event Stream Processing and machine learning algorithms can be used together to create a "digital twin" to monitor whether devices in disparate environments are operating properly and efficiently. Chapter 9 describes how SAS Event Stream Processing can detect changes in the quality of location-based data in real time and adjust it right away. Chapter 10 presents research using SAS Event Stream Processing for condition-based maintenance (CBM) of critical, high-valued machines. Chapter 11 explores how you can create intelligent computer vision systems, deploy those models on edge-devices to score streaming data, and use SAS Event Stream Processing to make decisions about what is seen in real time.

We Want to Hear from You

Do you have questions about a SAS Press book that you are reading? Contact us at saspress@sas.com.

SAS Press books are written *by* SAS Users *for* SAS Users. Please visit sas.com/books to sign up to request information on how to become a SAS Press author.

We welcome your participation in the development of new books and your feedback on SAS Press books that you are using. Please visit sas.com/books to sign up to review a book

Learn about new books and exclusive discounts. Sign up for our new books mailing list today at https://support.sas.com/en/books/subscribe-books.html.

Learn more about this author by visiting his author page at http://support.sas.com/harvey . There you can download free book excerpts, access example code and data, read the latest reviews, get updates, and more.

About the Author

 Michael Harvey is a Principal Technical Writer at SAS, serving as documentation project leader for the Internet of Things (IoT). Previously, Michael worked as a manager and a writer for EMC. He also teaches Information Architecture for the Duke Continuing Studies Technical Writing professional certificate program.

Michael has a BA in English and Psychology from the University of North Carolina at Chapel Hill and an MA in Experimental Psychology from Duke University. He has served in various leadership positions for the Carolina chapter of the Society for Technical Communication (STC) and has presented at local and international STC conferences. He was honored to be named an STC Fellow in 2011. As an instructor for the Durham Technical Community College Technical Writing program in the late 1980s and early 1990s, Michael worked to overhaul the curriculum, emphasizing the importance of developing technical curiosity and acquiring technical expertise.

Learn more about this author by visiting his author page at http://support.sas.com/harvey. There you can download free book excerpts, access example code and data, read the latest reviews, get updates, and more.

Chapter 1: Using SAS Event Stream Processing to Process Real World Events

By Michael Harvey, Robert Ligtenberg, and Jerry Baulier

Introduction

As Andrew G. Psaltis, the regional CTO for Cloudera, observes, "Data is flowing everywhere around us, through phones, credit cards, sensor-equipped buildings, vending machines, thermostats, trains, buses, planes, posts to social media, digital pictures and video – and the list goes on." Being able to harness that data presents abundant business opportunities. How can a business best capitalize on those opportunities?

The answer: SAS Event Stream Processing. It enables you to process and analyze continuously flowing real-world events in real time. Events arrive through high-throughput, low-latency data flows called event streams. These data flows are generated by occurrences such as sensor readings or market data. Each event within an event stream can be represented as a data record that consists of any number of fields. For example, an event generated by a pressure sensor could include two fields: a pressure reading and a timestamp. A more complex financial trade event could include multiple fields for transaction type, shares traded, price, broker, seller, stock symbol, timestamp, and so on. SAS Event Stream Processing can process the pressure data or the trades at any given moment. It can alert you to events of interest the instant that they occur.

Innovations in technology have enabled the reduction of the cost and size of sensors. Now sensors can be readily deployed within industrial equipment and consumer products. The number of sensors available has exploded, and a large portion of these sensors are now connected through the internet. The deluge of resulting data streams is often called *Big Data*. The Internet of Things (IoT) attaches a plethora of devices, sensors, and objects in our world to the internet. Big Data is collected and processed in real time from these "things."

SAS Event Stream Processing processes real-world data *as it is generated*. This instantly processed data is called *streaming* data. Processing streaming data introduces a paradigm shift from the traditional approach, where data is captured and stored in a database. After an event from an event stream is processed, it can be stored or discarded. Subsequent results of event stream data processing can also be stored and explored.

When time-sensitivity is important, processing streaming data at the point of generation is critical. For example, suppose that you are using sensing devices to track a customer who is browsing products at a retail establishment or online. Based on customer or product location (in real space or cyberspace), a system processing streaming data can generate an offer in real time to entice a purchase. An application that uses data at rest is not nimble enough to make these suggestions. Another example, also involving sensing devices, is the real-time tracking of vibrations in airliner engines. When anomalous patterns are detected (perhaps as the result of a bird impact), pilots can be alerted immediately so that they can take corrective action. Catastrophic failure can be avoided.

How Does SAS Event Stream Processing Work?

SAS Event Stream Processing reads from many source formats and outputs to many target formats (Figure 1.1).

Figure 1.1: Input and Output Streams Through the SAS Event Stream Processing Engine

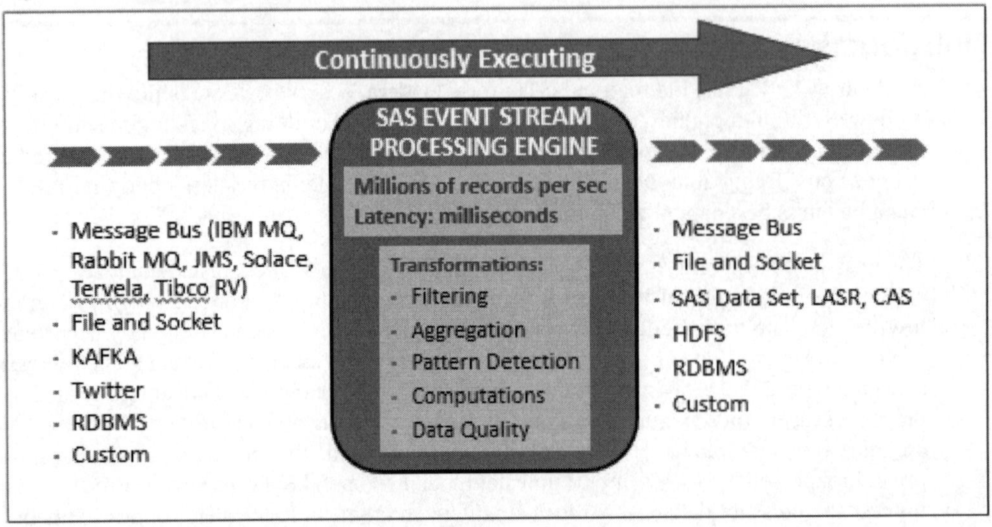

Many types of transformations are supported, including SQL primitives for filtering, aggregation, and pattern detection. There is built-in support for computations based on internal functions as well as on external languages like C++ and Python.

In addition, SAS Event Stream Processing provides many types of advanced analytical algorithms. These include algorithms for natural language text processing, image recognition and video image tracking, and machine learning. These analytical algorithms are covered in detail in Chapter 3.

In short, SAS Event Stream Processing provides a flexible platform with out-of-the-box capabilities to handle almost any type of event stream. You can use it to apply almost any type of business logic in real time.

What is a SAS Event Stream Processing Model?

SAS Event Stream Processing processes event streams through a *model*, which specifies how events are transformed and analyzed into meaningful results. The following figure (Figure 1.2) depicts the hierarchy of this model.

Figure 1.2: The SAS Event Stream Processing Model

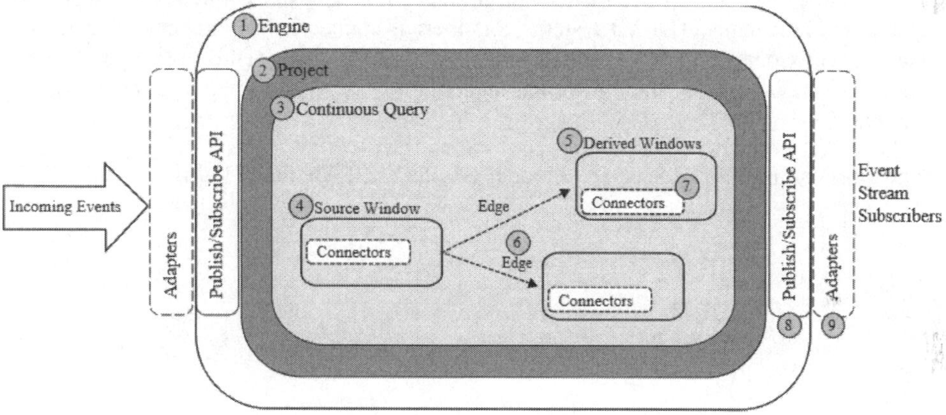

1. At the top of the model hierarchy is the *engine*. Each model contains only one engine instance with a unique name.
2. The engine contains one or more *projects*, each uniquely named. You can specify a port so that projects can be spread across network interfaces for throughput scalability.
3. A project contains one or more *continuous queries*. A continuous query is represented by a directed graph. This graph is a set of connected nodes that follow a direction down one or more parallel paths. Continuous queries are data flows, which are data transformations and analysis of incoming event streams.
4. Each query has a unique name and begins with one or more *Source windows*.
5. Source windows are typically connected to one or more *derived windows*. Derived windows can detect patterns in the data, transform the data, aggregate the data, analyze the data, or perform computations based on the data. They can be connected to other derived windows.
6. Windows are connected by *edges*, which have an associated direction.

7. *Connectors* are in-process to the engine. They use the publish/subscribe API to interface directly with a variety of message buses and brokers (for example, Kafka, RabbitMQ), communication fabrics, drivers, and clients.

8. The *publish/subscribe API* can be used to subscribe to an event stream window either from the same machine or from another machine on the network. Similarly, the publish/subscribe API can be used to publish event streams into a running event stream processing project Source window.

9. *Adapters* are stand-alone executable programs that can be networked. Some adapters are executable versions of their corresponding connector. Adapters use the publish/subscribe API to publish event streams to do the following:

    ○ publish event streams to Source windows

    ○ subscribe to event streams from any window

Connectors and adapters are available for almost any streaming format or protocol, including OPC-UA, Bacnet, PI System for sensors, as well as MQTT, RabbitMQ, Kafka, Tibco, Tervela, and IBM WebSphere for messaging fabrics. A database connector gives access to Database Management (DBM) systems, and there is connector support for sockets and websockets. Connectivity to Apache Camel and Apache NiFi extends the range of streaming sources even further. For more information about how connectors and adapters work, see Chapter 2.

The most common way to specify a model is with XML, as shown in Figure 1.3.

Figure 1.3: XML Code for a SAS Event Stream Processing Model

```
<engine name="engine" port="5555">
  <projects>
    <project name="project_A">

      <contqueries>
        <contquery name="contQuery">

          <windows>

            <window-source name="sourceWindow">
              <schema>
                <fields>
                  <field name="ID" type="int64" key="true"/>
                  <field name="symbol" type="string"/>
                  <field name="quantity" type="int32"/>
                  <field name="price" type="double"/>
                </fields>
              </schema>
              <connectors>
                <connector name="input" class="fs">
                  <properties>
                    <property name="type">pub</property>
                    <property name="fsname">localhost:4556</property>
                    <property name="fstype">csv</property>
                  </properties>
                </connector>
              </connectors>
            </window-source>

            <window-filter name="filterWindow">
              <expression>quantity > 1000</expression>
              <connectors>
                <connector name="output" class="fs">
                  <properties>
                    <property name="type">sub</property>
                    <property name="fsname">localhost:4557</property>
                    <property name="fstype">csv</property>
                  </properties>
                </connector>
              </connectors>
            </window-filter>

          </windows>

          <edges>
            <edge source="sourceWindow" target="filterWindow"/>
          </edges>

        </contquery>
      </contqueries>

    </project>
  </projects>
</engine>
```

This code reflects the hierarchy of a model. An engine contains a project, which contains a continuous query. Within the continuous query, the business logic is represented by a Source window and one derived window, which in this case is a Filter window. All event streams must enter continuous queries by being published or injected into a Source window. Event streams cannot be published or injected into any other window type. The Source window and Filter window in this model are connected by an edge, as specified in the **<edge>** XML element.

The Source window is configured to ingest events from an event stream through a port on a server. The server in this example is the localhost server. The Source window also specifies a schema that defines field properties for the incoming events. In this case, there are four fields: an ID that serves as the event's key field, a stock symbol, a quantity (shares transacted), and a price.

The Filter window is configured to select events that satisfy a specific condition. In this example, the condition requires that the quantity is greater than 1000. Satisfying events are passed on while non-satisfying events are not passed on. A subscribing connector passes the output events to a port where another application might be listening.

You deploy models by executing them within a SAS Event Stream Processing engine. Upon execution, the engine establishes a connection to the specified input port and starts ingesting incoming events. The events are processed according to the configured business logic and output events are delivered to the output port.

Processing Events in Derived Windows

All continuous queries contain one or more Source windows and one or more derived windows. SAS Event Stream Processing supports a variety of derived window types, each having a specialized purpose (Table 1.1).

Table 1.1: Derived Window Types

Window Type	Description
Aggregate	Calculates aggregates like sum and average, similar to SQL group-by aggregations.
Compute	Adds calculated fields to an event stream.
Copy	Copies (replicates) an event stream and supports a retention policy. Retention is used for time-based or count-based windowing.
Counter	Counts events and calculates throughput.
Filter	Selects events based on an expression.
Functional	Executes user-defined functions. Supports regular expressions and looping to parse XML and JSON fields.
Geofence	Determines whether event locations are in or near an area of interest.
Join	Joins two event streams like an SQL join. Supports inner, left-outer, right-outer, and full outer joins.
Notification	Sends notifications. Supports SMTP, SMS, and MMS.
Object Tracking	Performs multi-object tracking (MOT) in real time.
Pattern	Detects patterns and anomalies within events and across events.
Procedural	Calls external functions (C++, SAS). Supports multiple input windows (input streams) with an input-handler function for each input.
Remove State	Facilitates the transition of a stateful part of a model to a stateless part of a model.
Text Category	Categorizes text fields.
Text Context	Extracts text and classifies terms.
Text Sentiment	Performs sentiment analysis.

Window Type	Description
Text Topic	Scores and identifies themes.
Transpose	Transposes rows to columns or columns to rows.
Union	Combines multiple event streams with the same schema into a single stream, like an SQL union.

Each of these windows is explained in detail in the "Using Source and Derived Windows" documentation, which is available at https://go.documentation.sas.com/?cdcId=espcdc&cdcVersion=6.2&docsetId=espcreatewind ows&docsetTarget=titlepage.htm&locale=en.

Examples of Event Transformations

The following two examples demonstrate how events are transformed within a continuous query. The first uses a Join window to combine data from two separate event streams into a single output stream. The second combines a Pattern window and Notification window to catch front-running trades.

Example: Using a Join Window

A Join window combines fields from two input event streams into a single output event stream. In the model depicted in Figure 1.4, the Join window (AddTraderName) receives events from a left input window (a Source window named Traders) and a right input window (a Filter window named LargeTrades). It produces a single output stream of joined events. Joined events are created according to a user-specified join type and user-defined join conditions.

Figure 1.4: Continuous Query That Uses a Simple Join

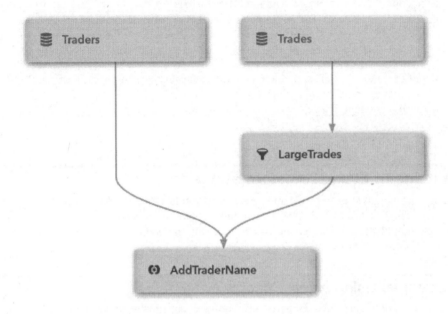

The Source window named Trades streams data about securities transactions. A file and socket connector is established to publish events into the Trades window from a file named trades.csv in the /data directory.

Example Code 1.1

```
<window-source name='Trades' index='pi_RBTREE'>
      <schema>
          <fields>
             <field name='tradeID' type='string' key='true'/>
             <field name='security' type='string'/>
             <field name='quantity' type='int32'/>
             <field name='price' type='double'/>
             <field name='traderID' type='int64'/>
             <field name='time' type='stamp'/>
          </fields>
      </schema>
      <connectors>
          <connector class="fs" name="publisher">
            <properties>
              <property name="type">pub</property>
              <property name="fstype">csv</property>
              <property name="fsname">/data/trades.csv</property>
              <property name="transactional">true</property>
              <property name="blocksize">1</property>
              <property name="dateformat">
              %d/%b/%Y:%H:%M:%S
              </property>
            </properties>
          </connector>
```

```
        </connectors>
    </window-source>
</windows>
```

A Filter window named LargeTrades receives events from the Trades window. It filters out any event that involves fewer than 100 shares.

Example Code 1.2

```
<window-filter name='LargeTrades'>
      <expression>quantity >= 100</expression>
</window-filter>
```

A second Source window named Traders streams data about who performs those transactions. A file and socket connector is established to publish events into the Traders window from a file named traders.csv in the /data directory.

Example Code 1.3

```
<window-source name='Traders'>
     <schema>
        <fields>
            <field name='ID' type='int64' key='true'/>
            <field name='name' type='string'/>
        </fields>
     </schema>
     <connectors>
            <connector class="fs" name="publisher">
              <properties>
                <property name="type">pub</property>
                <property name="fstype">csv</property>
                <property name="fsname">/data/traders.csv</property>
                <property name="transactional">true</property>
                <property name="blocksize">1</property>
              </properties>
            </connector>
        </connectors>
</window-source>
```

The Join window named AddTraderName matches filtered transactions from the first Source window with their associated traders from the second. Specifically, the conditions tag matches the traderID values specified in the transactions Source window to the ID values specified in the traders Source window.

Example Code 1.4

```
<window-join name='AddTraderName'>
      <join type="leftouter">
         <conditions>
             <fields left='traderID' right='ID' />
         </conditions>
      </join>
      <output>
         <field-selection name='security' source='l_security'/>
         <field-selection name='quantity' source='l_quantity'/>
         <field-selection name='price' source='l_price'/>
         <field-selection name='traderID' source='l_traderID'/>
         <field-selection name='time' source='l_time'/>
```

```
            <field-selection name='name' source='r_name'/>
        </output>
</window-join>
```

By default, this join order is determined through the specification of edges. The left window is the first window that is defined as a connecting edge to the Join window. The right window is the second window that is defined as a connecting edge.

Example Code 1.5

```
                                <edges>
        <edge source="LargeTrades" target="AddTraderName"/>
        <edge source='Traders' target='AddTraderName'/>
        <edge source='Trades' target='LargeTrades'/>
</edges>
```

Suppose that the following events stream through the Source window named Trades (Figure 1.5):

Figure 1.5: Input to the Trades Source Window

Opcode	tradeID	security	quantity	price	traderID	time
Insert	TID1234331	ibm	80	100.300000	10004	2012-07-08T08:1...
Insert	TID1234330	sap	90	-32.000000	10003	2012-07-08T08:1...
Insert	TID1234329	ibm	1000	100.000000	10004	2012-07-08T08:1...
Insert	TID1234328	sap	1000	32.000000	10003	2012-07-08T08:1...
Insert	TID1234327	ibm	1000	100.300000	10002	2012-07-08T08:1...
Insert	TID1234326	sap	1000	34.300000	10003	2012-07-08T08:1...
Insert	TID1234325	ibm	1000	100.400000	10004	2012-07-08T08:1...
Insert	TID1234324	ibm	1000	100.300000	10004	2012-07-08T08:1...
Insert	TID1234323	ibm	1000	100.200000	10004	2012-07-08T08:1...
Insert	TID1234322	sap	750	34.200000	10003	2012-07-08T08:1...
Insert	TID1234321	ibm	1000	100.100000	10002	2012-07-08T08:1...

Figure 1.6 shows the events that stream from the Filter window.

Figure 1.6: Output from the LargeTrades Filter Window

Opcode	tradeID	security	quantity	price	traderID	time
Insert	TID1234329	ibm	1000	100.000000	10004	2012-07-08T08:1...
Insert	TID1234328	sap	1000	32.000000	10003	2012-07-08T08:1...
Insert	TID1234327	ibm	1000	100.300000	10002	2012-07-08T08:1...
Insert	TID1234326	sap	1000	34.300000	10003	2012-07-08T08:1...
Insert	TID1234325	ibm	1000	100.400000	10004	2012-07-08T08:1...
Insert	TID1234324	ibm	1000	100.300000	10004	2012-07-08T08:1...
Insert	TID1234323	ibm	1000	100.200000	10004	2012-07-08T08:1...
Insert	TID1234322	sap	750	34.200000	10003	2012-07-08T08:1...
Insert	TID1234321	ibm	1000	100.100000	10002	2012-07-08T08:1...

Now suppose that these events stream through the Traders Source window (Figure 1.7).

Figure 1.7: Input into the Traders Source Window

Opcode	ID	name
Insert	10004	dan penteado
Insert	10003	bernie
Insert	10002	steals a. lot

Figure 1.8 is the result of matching filtered transactions from the first Source window with their associated traders from the second.

Figure 1.8: Output from the AddTraderName Join Window

Opcode	tradeID	security	quantity	price	traderID	time	name
Insert	TID1234329	ibm	1000	100.000000	10004	2012-07-08T0...	dan penteado
Insert	TID1234328	sap	1000	32.000000	10003	2012-07-08T0...	bernie
Insert	TID1234327	ibm	1000	100.300000	10002	2012-07-08T0...	steals a. lot
Insert	TID1234326	sap	1000	34.300000	10003	2012-07-08T0...	bernie
Insert	TID1234325	ibm	1000	100.400000	10004	2012-07-08T0...	dan penteado
Insert	TID1234324	ibm	1000	100.300000	10004	2012-07-08T0...	dan penteado
Insert	TID1234323	ibm	1000	100.200000	10004	2012-07-08T0...	dan penteado
Insert	TID1234322	sap	750	34.200000	10003	2012-07-08T0...	bernie
Insert	TID1234321	ibm	1000	100.100000	10002	2012-07-08T0...	steals a. lot

Example: Using a Pattern Window and a Notification Window

A Pattern window enables you to detect events of interest (EOIs) as they stream through. To create a Pattern window, specify a list of EOIs and assemble them into an expression that uses logical operators. A Notification window enables you to send notifications through email using the Simple Mail Transfer Protocol (SMTP), text using the Short Message Service (SMS), or send multimedia messages using the Multimedia Messaging Service (MMS).

The following example uses a Pattern window to catch stock traders when they attempt front-running buys. A broker caught in the act is sent an email, an SMS text message, and an MMS message produced by a Notification window (Figure 1.9).

Figure 1.9: Continuous Query to Detect Front-Running Buys

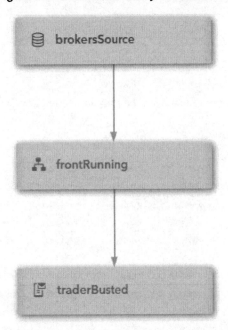

The message sent includes details of the trades involved in the violation, and for the channels that permit graphics, the message also contains an image of someone in a jail cell. All relevant message routing information is included in the broker dimension data streamed into the Source window.

Example Code 1.6

```
i,n,1012112,Frodo,ESP,940 Orion Suite 201 Cary NC
27513,,frodo.doe@orion.com,5556466705,txt.att.net,mms.att.net
i,n,1012223,Sam,ESP,940 Orion Suite 201 Cary NC
27513,,sam.doe@orion.com,5556466706,txt.att.net,mms.att.net
i,n,1012445,Pippin,ESP,940 Orion Suite 201 Cary NC
27513,pippin.doe@orion.com,5556466707,txt.att.net,mms.att.net
i,n,1012334,Merry,ESP,940 Orion Suite 201 Cary NC
27513,merry.doe@orion.com,5556466708,txt.att.net,mms.att.net
i,n,101667,Gandalf,ESP,940 Orion Suite 201 Cary NC
27513,gandalf.doe@orion.com,5556466709,txt.att.net,mms.att.net
i,n,1012001,Aragorn,ESP,940 Orion Suite 201 Cary NC
27513,aragorn.doe@orion.com,5556466710,txt.att.net,mms.att.net
```

Note that the last four fields contain the email, phone number, and SMS and MMS gateways for each broker.

First, data streams into the model through a Source window named brokersSource.

Example Code 1.7

```
<window-source name='brokersSource' insert-only='true'>
   <schema-string>broker*:int32,brokerName:string,brokerage:string,
```

```
brokerAddress:string,brokerPhone:string,email:string,
                smsGateway:string,mmsGateway:string</schema-string>
    <connectors>
        <connector class='fs'>
            <properties>
                <property name='type'>pub</property>
                <property name='fstype'>csv</property>
                <property name='fsname'>data/brokers.csv</property>
            </properties>
        </connector>
    </connectors>
</window-source>
```

The Pattern window is constructed to detect front-running violations. The window deals with up to three events (trades) at a time (e1, e2, and e3). Each trade contains broker and customer information as well as the trade data. All data must be available in order to format a notification message.

The Pattern window looks like Example Code 1.8.

Example Code 1.8

```
<window-pattern name='frontRunning'>
    <schema>
        <fields>
            <field name='id' type='int64' key='true' />
            <field name='broker' type='int32' />
            <field name='brokerName' type='string' />
            <field name='email' type='string' />
            <field name='phone' type='string' />
            <field name='sms' type='string' />
            <field name='mms' type='string' />
            <field name='customer' type='int32' />
            <field name='symbol' type='string' />
            <field name='tstamp1' type='string' />
            <field name='tstamp2' type='string' />
            <field name='tstamp3' type='string' />
            <field name='tradeId1' type='int32' />
            <field name='tradeId2' type='int32' />
            <field name='tradeId3' type='int32' />
            <field name='price1' type='double' />
            <field name='price2' type='double' />
            <field name='price3' type='double' />
            <field name='quant1' type='int32' />
            <field name='quant2' type='int32' />
            <field name='quant3' type='int32' />
            <field name='slot' type='int32' />
        </fields>
    </schema>
    <splitter-expr>
        <expression>slot</expression>
    </splitter-expr>
    <patterns>
        <pattern index='broker,symbol'>
            <events>
                <event name='e1'>((buysellflg == 1)
                  and (broker == buyer)
```

```
      and (s == symbol)
      and (b == broker)
      and (p == price))</event>
   <event name='e2'>((buysellflg == 1)
    and (broker != buyer)
    and (s == symbol)
    and (b == broker))</event>
   <event name='e3'><![CDATA[((buysellflg == 0)
    and (broker == seller)
    and (s == symbol)
    and (b == broker)
    and (p < price))]]></event>
  </events>
  <logic>fby{1 hour}(fby{1 hour}(e1,e2),e3)</logic>
  <output>
     <field-selection name='broker' node='e1'/>
     <field-selection name='brokerName' node='e1'/>
     <field-selection name='brokerEmail' node='e1'/>
     <field-selection name='brokerPhone' node='e1'/>
     <field-selection name='brokerSms' node='e1'/>
     <field-selection name='brokerMms' node='e1'/>
     <field-selection name='buyer' node='e2'/>
     <field-selection name='symbol' node='e1'/>
     <field-selection name='date' node='e1'/>
     <field-selection name='date' node='e2'/>
     <field-selection name='date' node='e3'/>
     <field-selection name='id' node='e1'/>
     <field-selection name='id' node='e2'/>
     <field-selection name='id' node='e3'/>
     <field-selection name='price' node='e1'/>
     <field-selection name='price' node='e2'/>
     <field-selection name='price' node='e3'/>
     <field-selection name='quant' node='e1'/>
     <field-selection name='quant' node='e2'/>
     <field-selection name='quant' node='e3'/>
     <field-expr>1</field-expr>
  </output>
</pattern>
<pattern index='broker,symbol'>
  <events>
     <event name='e1'>((buysellflg == 0)
      and (broker == seller)
      and (s == symbol)
      and (b == broker))</event>
     <event name='e2'>((buysellflg == 0)
      and (broker != seller)
      and (s == symbol)
      and (b == broker))</event>
  </events>
  <logic>fby{10 minutes}(e1,e2)</logic>
  <output>
     <field-selection name='broker' node='e1'/>
     <field-selection name='brokerName' node='e1'/>
     <field-selection name='brokerEmail' node='e1'/>
     <field-selection name='brokerPhone' node='e1'/>
     <field-selection name='brokerSms' node='e1'/>
     <field-selection name='brokerMms' node='e1'/>
     <field-selection name='seller' node='e2'/>
```

```
                    <field-selection name='symbol' node='e1'/>
                    <field-selection name='date' node='e1'/>
                    <field-selection name='date' node='e2'/>
                    <field-expr> </field-expr>
                    <field-selection name='id' node='e1'/>
                    <field-selection name='id' node='e2'/>
                    <field-expr>0</field-expr>
                    <field-selection name='price' node='e1'/>
                    <field-selection name='price' node='e2'/>
                    <field-expr>0</field-expr>
                    <field-selection name='quant' node='e1'/>
                    <field-selection name='quant' node='e2'/>
                    <field-expr>0</field-expr>
                    <field-expr>2</field-expr>
                </output>
            </pattern>
        </patterns>
</window-pattern>
```

Events stream from the Pattern window into the Notification window.

Example Code 1.9

```
<window-notification name='traderBusted'>
    <smtp host='smtp-server.ec.rr.com'
          user='esptest@ec.rr.com'
          password='esptest1' port='587' />
    <schema>
        <fields>
            <field name='id' type='int64' key='true' />
            <field name='broker' type='int32' />
            <field name='brokerName' type='string' />
            <field name='email' type='string' />
            <field name='phone' type='string' />
            <field name='sms' type='string' />
            <field name='mms' type='string' />
            <field name='customer' type='int32' />
            <field name='symbol' type='string' />
            <field name='tstamp1' type='string' />
            <field name='tstamp2' type='string' />
            <field name='tstamp3' type='string' />
            <field name='tradeId1' type='int32' />
            <field name='tradeId2' type='int32' />
            <field name='tradeId3' type='int32' />
            <field name='price1' type='double' />
            <field name='price2' type='double' />
            <field name='price3' type='double' />
            <field name='quant1' type='int32' />
            <field name='quant2' type='int32' />
            <field name='quant3' type='int32' />
            <field name='slot' type='int32' />
            <field name='day' type='string' />
            <field name='price1' type='double' />
            <field name='price2' type='double' />
            <field name='price3' type='double' />
            <field name='time1' type='string' />
            <field name='time2' type='string' />
            <field name='time3' type='string' />
```

```
            <field name='profit' type='double' />
        </fields>
    </schema>
    <function-context>
        <properties>
            <property-list name='time1' delimiter=' '>$tstamp1
            </property-list>
            <property-list name='time2' delimiter=' '>$tstamp2
            </property-list>
            <property-list name='time3' delimiter=' '>$tstamp3
            </property-list>
        </properties>
        <functions>
            <function name='profit'>
             product($quant3,diff($price3,$price1))</function>
            <function name='day'>listItem(#time1,0)</function>
            <function name='time1'>listItem(#time1,1)</function>
            <function name='time2'>listItem(#time2,1)</function>
            <function name='time3'>listItem(#time3,1)</function>
            <function name='price1'>precision($price1,2)</function>
            <function name='price2'>precision($price2,2)</function>
            <function name='price3'>precision($price3,2)</function>
        </functions>
    </function-context>
    <delivery-channels>
        <email test='true' throttle-interval='1 day'>
            <deliver>contains(toLower($brokerName),'@BROKER@')</deliver>
            <email-info>
                <sender>esptest@ec.rr.com</sender>
                <recipients>$email</recipients>
                <from>ESP Broker Surveillance</from>
                <to>$brokerName</to>
                <subject>You have been caught cheating,
                  $brokerName</subject>
            </email-info>
            <email-contents>
                <html-content><![CDATA[
                <body>You bought <b>$quant1</b> shares of <b>$symbol</b>
                     for $<b>$price1</b> on <b>$day</b> at <b>$time1</b>.
                     You then bought <b>$symbol</b> for customer
                     <b>$customer</b>
                     at <b>$time2</b>, after which you sold
                     <b>$quant3</b> shares of
                     <b>$symbol</b> at <b>$time3</b> for $<b>$price3</b>,
                     thus making you a profit of $<b>$profit</b>.
                     <br/><br/></body>
                ]]></html-content>
                <image-content type='image'>
                 http://esp-base:18080/esp/stuff/jail.jpg
                </image-content>
            </email-contents>
        </email>
        <mms test='true' throttle-interval='1 day'>
            <deliver>contains(toLower($brokerName),'@BROKER@')</deliver>
            <mms-info>
                <sender>esptest@ec.rr.com</sender>
                <subject>You have been caught cheating,
                  $brokerName</subject>
```

```
            <gateway>$mms</gateway>
            <phone>$phone</phone>
        </mms-info>
        <mms-contents>
            <text-content>You bought $quant1 shares of $symbol for
            $$price1 on $day at $time1. You then bought $symbol for
            customer $customer at $time2, after which you sold $quant3
            shares of $symbol at $time3 for $$price3, thus making you
            a profit of $$profit.
            </text-content>
            <image-content type='image'>
             http://esp-base:18080/esp/stuff/x.jpg
            </image-content>
        </mms-contents>
    </mms>
    <sms test='true' throttle-interval='1 day'>
        <deliver>contains(toLower($brokerName),'@BROKER@')</deliver>
        <sms-info>
            <sender>esptest@ec.rr.com</sender>
            <subject>You have been caught, $brokerName</subject>
            <from>ESP Broker Surveillance</from>
            <gateway>$sms</gateway>
            <phone>$phone</phone>
        </sms-info>
        <sms-contents>
            <text-content>You bought $quant1 shares of $symbol
            for $$price1 on $day at $time1. You then bought $symbol
            for customer $customer at $time2, after which you sold
            $quant3 shares of $symbol at $time3 for $$price3,
            thus making you a profit of $$profit.</text-content>
        </sms-contents>
    </sms>
  </delivery-channels>
</window-notification>
```

Because this example uses MMS, you need to define a different SMTP server. Any email account referenced by that server must be specified in your SMTP configuration. The window calculates fields to use when formatting notification messages to the broker. A schema and a function context are defined.

When an event comes in, functions are run on the input event and schema data is created. You can use values from either the input event or the schema data in the message content. For example, see Example Code 1.10.

Example Code 1.10

```
We noticed you bought <b>$quant1</b> shares of <b>$symbol</b> for
$<b>$price1</b>
on <b>$day</b> at <b>$time1</b>. You then bought <b>$symbol</b> for
customer <b>$customer</b> at <b>$time2</b>, after which
you sold <b>$quant3</b> shares of <b>$symbol</b> at <b>$time3</b>
for $<b>$price3</b>, thus making you a profit of $<b>$profit</b>.
```

Note the number of variable references. Some of the variable references are to the schema data (quant1, price1, price3, ...), and some to the input data (symbol). Variable references are also used to resolve the routing information for the notification.

Example Code 1.11

```
<recipients>$email</recipients>
<gateway>$sms</gateway>
<phone>$phone</phone>
```

A function is used to determine when to send the notification. The same deliver function is used for all channels.

Example Code 1.12

```
<deliver>contains(toLower($brokerName),'@BROKER@')</deliver>
```

Whenever you see the notation @TOKEN@ in an XML model, this means that the token is resolved when the project is loaded. These tokens can be resolved in one of three ways:

- on the command line, for example, dfesp_xml_server -BROKER pippin
- in your environment, for example, $ export BROKER=pippin
- in the properties for a project, for example, <property name='BROKER'>pippin</property>

In this case, you can specify which broker to use to send a notification.

Streaming Analytics

SAS provides an extensive toolset of analytical algorithms and machine learning techniques, which can be directly applied to real-time streams in SAS Event Stream Processing models. You can use algorithms and techniques to address the following common challenges with data from the Internet of Things (IoT):

- lots of disparate variables
- noisy or missing data
- redundancy in the data
- prediction of rare events

Common use cases for streaming analytics include the following:

- preprocessing, transforming, or filtering data – determining how much and what data to send from the edge to the data center
- detecting anomalies
- monitoring system stability or degradation
- processing unstructured text, audio, video, or image data in order to discern patterns or trends

You can take one of two approaches with streaming analytics:

1. Use collected and stored data to develop analytical models off-line and then deploy them to SAS Event Stream Processing models to analyze and score incoming streams in real time.
2. Train analytical models in SAS Event Stream Processing on incoming streams and then update the models online based on the latest training results.

Additional derived window types are provided to handle analytical algorithms and machine learning techniques (Table 1.2).

Table 1.2: Analytics Window Types

Analytics Window Type	Description
Calculate	Applies a variety of analytical algorithms to event streams.
Model Reader	Accepts analytical models that were developed off-line.
Model Supervisor	Manages executing analytical models based on update requests.
Train	Uses incoming events to refine analytical model parameters.
Score	Scores incoming events based on the latest values of the analytical model parameters.

Chapter 3 reviews the innovations in advanced analytical models that are trained on data at rest and scored on streaming data, along with those that are directly applied to streaming data.

Using SAS Micro Analytic Service Modules with Streaming Analytics

A SAS Micro Analytic Service (MAS) module is essentially a named block of code that you execute within a model. This block, which you define at the project level, can contain one or more functions. You define a MAS map in a Calculate window to bind a function to any of its input windows. This binding acts as the input handler for the Calculate window.

Consider the following continuous query (Figure 1.10):

Figure 1.10: Continuous Query Containing a Source Window Streaming into a Calculate Window

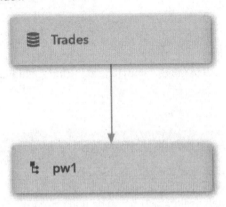

The Source window Trades contains a schema that specifies the fields that define the structure of incoming events. Events stream into the Source window through a file and socket publisher connector. The input data is contained in a file named input.csv in the current working directory.

Example Code 1.13

```
<window-source name='Trades' index='pi_RBTREE'>
        <schema>
          <fields>
            <field name='tradeID' type='string' key='true'/>
            <field name='security' type='string'/>
            <field name='quantity' type='int32'/>
            <field name='price' type='double'/>
            <field name='traderID' type='int64'/>
            <field name='time' type='string'/>
          </fields>
        </schema>
        <connectors>
          <connector class='fs' name='pub'>
            <properties>
              <property name='type'>pub</property>
              <property name='fstype'>csv</property>
              <property name='blocksize'>1</property>
            </properties>
          </connector>
        </connectors>
      </window-source>
```

A MAS module named module_1 is defined at the project level. It contains a function named compute_total written in Python. The Python code for compute_total is specified within the <code> element of the module. This function acts as the input handler for all events that are passed from the input window to the Calculate window.

Example Code 1.14

```
<mas-modules>
    <mas-module language="python" module="module_1" func-
names='compute_total' >
      <code>
        <![CDATA[
        def compute_total(quantity, price):
            "Output: total"
            total = quantity * price
        return total
        ]]>
      </code>
</mas-module>
  </mas-modules>
```

Data relevant to the security being traded, the quantity of shares, the current prices, and the person who performed the trade are streamed into the Calculate window named pw1. (Before SAS Event Stream Processing 5.2, support for MAS modules was provided through the Procedural window.) At the map level of the Calculate window, the window map binds the compute_total function of module_1 to the input window.

Example Code 1.15

```
<window-calculate name='pw1' algorithm='MAS'>
        <schema>
          <fields>
            <field name='tradeID' type='string' key='true'/>
            <field name='security' type='string'/>
            <field name='quantity' type='int32'/>
            <field name='price' type='double'/>
            <field name='traderID' type='int64'/>
            <field name='time' type='string'/>
            <field name='total' type='double'/>
          </fields>
        </schema>
        <mas-map>
          <window-map module="module_1" revision="0" source="Trades"
function="compute_total"/>
        </mas-map>
  </window-calculate>
```

Edges connect the Source window to the Calculate window.

Example Code 1.16

```
<edges>
        <edge source='Trades' target='pw1' role='data'/>
</edges>
```

Suppose that you stream the following events through the Source window (Figure 1.11):

Figure 1.11: Events That Stream Through the Trades Window

Opcode	tradeID	security	quantity	price	traderID	time
Insert	TID1234329	ibm	1000	100.000000	10004	08/Jul/2012:08:1...
Insert	TID1234328	sap	1000	32.000000	10003	08/Jul/2012:08:1...
Insert	TID1234327	ibm	1000	100.300000	10002	08/Jul/2012:08:1...
Insert	TID1234326	sap	1000	34.300000	10003	08/Jul/2012:08:1...
Insert	TID1234325	ibm	1000	100.400000	10004	08/Jul/2012:08:1...
Insert	TID1234324	ibm	1000	100.300000	10004	08/Jul/2012:08:1...
Insert	TID1234323	ibm	1000	100.200000	10004	08/Jul/2012:08:1...
Insert	TID1234322	sap	750	34.200000	10003	08/Jul/2012:08:1...
Insert	TID1234321	ibm	1000	100.100000	10002	08/Jul/2012:08:1...

Figure 1.12 shows the resulting events from the Calculate window.

Figure 1.12: Results From the Calculate Window

Opcode	tradeID	security	quantity	price	traderID	time	total
Insert	TID1234329	ibm	1000	100.000000	10004	08/Jul/2012:0...	100,000.000000
Insert	TID1234328	sap	1000	32.000000	10003	08/Jul/2012:0...	32,000.000000
Insert	TID1234327	ibm	1000	100.300000	10002	08/Jul/2012:0...	100,300.000000
Insert	TID1234326	sap	1000	34.300000	10003	08/Jul/2012:0...	34,300.000000
Insert	TID1234325	ibm	1000	100.400000	10004	08/Jul/2012:0...	100,400.000000
Insert	TID1234324	ibm	1000	100.300000	10004	08/Jul/2012:0...	100,300.000000
Insert	TID1234323	ibm	1000	100.200000	10004	08/Jul/2012:0...	100,200.000000
Insert	TID1234322	sap	750	34.200000	10003	08/Jul/2012:0...	25,650.000000
Insert	TID1234321	ibm	1000	100.100000	10002	08/Jul/2012:0...	100,100.000000

For more information about using MAS modules in Calculate windows, see "Working with SAS Micro Analytic Service Modules" documentation at https://go.documentation.sas.com/?cdcId=espcdc&cdcVersion=6.2&docsetId=espan&docsetTarget=p1qv3axlms1ckmn1gv3hub74xmue.htm&locale=en.

Addressing Big Data and the Internet of Things

SAS Event Stream Processing enables solutions to meet IoT challenges through reference architectures, which are discussed in detail in Chapter 5. Figure 1.13 shows a common reference architecture.

Figure 1.13: Internet of Things Reference Architecture

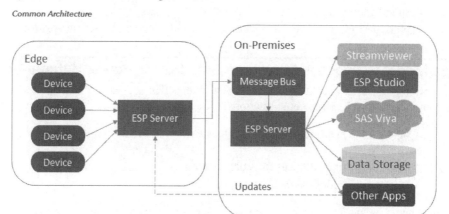

Here, event streams from sensor devices are ingested by an ESP server running on an edge device. The edge device could be on an airliner, a tanker, or a locomotive. There could be any number of these edge devices in operation. The edge ESP servers can be running ESP models that are configured to detect worrisome combinations of sensor values. Specific events of interest trigger alerts to the operator of the edge device, which are then communicated through a message bus to the on-premises ESP server.

The on-premises ESP server aggregates events that are received from the edge devices and performs further analytical model refinement as well as analysis, monitoring, and reporting functions. This ESP server can send analyzed events to a data storage device or to other applications for further processing. SAS Event Stream Processing Streamviewer can be used to visualize events as they stream through the ESP server. SAS Event Stream Processing Studio can interact with the ESP server to create, edit, upload, publish, and test event stream processing models. This enables ongoing streaming project development and analytical model development. The on-premises ESP server can also interact with a SAS Cloud Analytic Services (CAS) server, which is shipped with SAS Viya.

This common reference architecture can be applied to any number of real-world use cases. Consider the analysis of power transmission on a grid that is monitored with Phasor Measurement Units (PMUs). PMUs take high-frequency measurements of power frequency, voltage, current, and phase angle at different locations along the grid. They use GPS signaling to ensure the time accuracy of measurements taken at different locations. High-frequency and time-accurate measurements enable operators and engineers to make informed decisions as they monitor and control the grid.

To maintain stability on a power grid, operators and engineers can use this reference architecture to do the following:

- Understand the steady state operation of the grid using streaming analytics that calculate descriptive statistics from real-time and historical PMU measurement data.
- Detect anomalous events on the grid in real time by monitoring PMUs and comparing measurements with steady state descriptive statistics.

- Categorize events by type, count, intensity, time, location, and equipment type.
- Respond appropriately to detected events.
- Capture data for post-event analysis using a high-volume data storage platform.

Edge Model to Process Measurements from a Power Substation

In the power-grid use case, the edge model ingests events that consist of PMU measurement data from various locations (power substations). The event schema includes the following:

- A field for the measurement type, such as voltage, current, frequency, and angle measurement.
- A field for the time of the event's observation.
- A field that specifies the measurement's value.

The edge model includes a Source window and a series of derived windows to detect events of interest and perform analytical calculations on the PMU data. The model calculates and compares forecasted data with observed data with a series of Aggregate, Compute, Copy, Counter, Functional, and Join windows. A downstream Compute window calculates the control limits for each PMU based on derived statistics. After the control limit data is calculated, it is published from the Compute window to a Kafka broker through a Kafka subscriber connector.

On-Premises Model for Further Processing

The on-premises model (Figure 1.14) reads data from the Kafka broker that receives data from the ESP servers on the edge.

Figure 1.14: On-Premises Model

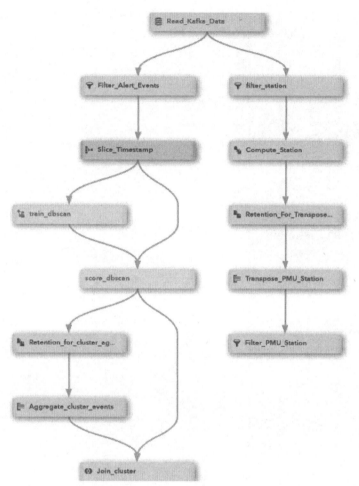

A Source window in the on-premises model receives data from the Kafka broker using a Kafka publisher connector. The schema of the Source window on-premises is similar to the schema of the Compute window that published to the Kafka broker from the edge. All the fields are the same, but here, the Source window has a key field that is unique to the location (substation) and measurement type rather than a key field unique to just the measurement type. From the Source window, the model splits the data stream into two branches:

The first branch combines incoming data events with unique measurement types and locations for a given timestamp into a single event for each timestamp with fields for measurement values of each type and location (Figure 1.15).

Figure 1.15: First Branch of On-Premises Model

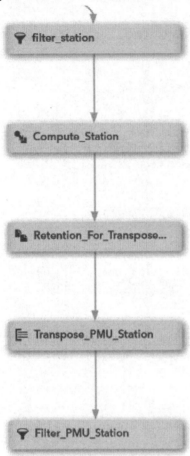

A Compute window creates an event where the measurement value field is the value for the location and measurement type corresponding to the station field. The window places a null value in the measurement value fields of all locations and measurement types that do not correspond to the station field.

An Aggregate window downstream of the Compute window populates null event fields with the last non-null value affecting that field within each 30-second interval. A CAS adapter subscribes to the events streaming through the Aggregate window in order to store the event data for further processing.

The second branch sends data events with values outside the upper and lower control limits to a training and scoring clustering model that is used to detect events of interest across the power grid (Figure 1.16).

Figure 1.16: Second Branch of On-Premises Model

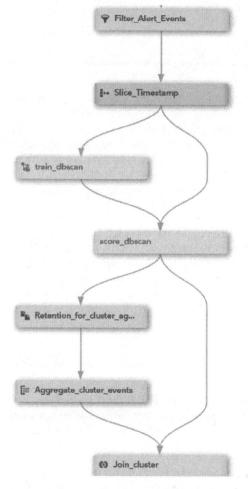

A Filter window filters events with values outside the upper and lower control limits. Values outside the control limits are considered alerts.

A Procedural window calculates geographical information for the stations in the event and creates new fields for the ZIP code, longitude, and latitude of each event's station.

A Train window downstream of the Procedural window clusters the events based on the timestamp, using the DBSCAN clustering algorithm. A Score window receives data events from the Procedural window and model events from the Train window. The Score window scores the data events using the trained DBSCAN mode. It assigns a cluster ID to each event based on its timestamp, along with the minimum distance of the event to the cluster centroid.

An Aggregate window downstream of the Score window counts the events in each cluster for each minute interval. A Join window combines the events that stream through the Aggregate window with the events from the Score window by cluster ID and ZIP code.

Two SAS Cloud Analytic Server (CAS) subscriber adapters read in data events from the two branches of the model. After that data has been loaded to CAS tables, you can interact with snapshots of the data streams with SAS Visual Analytics.

Conclusion

SAS Event Stream Processing is an enterprise-class application designed to address the Big Data and IoT challenges faced by most modern businesses. It provides the flexibility to build event stream processing models that ingest event streams from any source and apply business logic of any type and complexity. You can use it to analyze structured and unstructured data sources, including video, text, and image classification and identification, using advanced analytics with embedded AI and machine learning capabilities. As a component of the wider SAS software suite of applications, it can tap directly into the proven capabilities around data integration, data quality, and advanced analytics.

Returning to the observations of Andrew Psaltis, "[T]he digital universe is doubling in size every two years. ...A great way of putting that in perspective [is to realize that] if a byte of data were a gallon of water, in only 10 seconds there would be enough data to fill an average home. In 2020, it will only take 2 seconds." SAS Event Stream Processing has the capability to process and analyze those streams no matter how quickly the digital universe grows. You will never be underwater when you use it to capitalize on the opportunities presented by the Internet of Things.

About the Contributors

As Principal Technical Training Consultant in SAS Education, **Robert Ligtenberg** develops and delivers customer training courses in the Data Management space. Robert has a PhD in physics from North Carolina State University and an undergraduate degree from the University of Twente, the Netherlands.

As the Vice President of Research & Development for all Internet of Things (IOT) offerings at SAS, **Jerry Baulier** works closely with customers, partners, and industry analysts to help research and development teams at SAS develop IOT vertical solutions on the Event Stream Processing product suite, Analytics for IOT product suite, and the Quality Analytics product suite. Jerry holds a master's degree from Stevens Technical Institute and a bachelor's degree from UMAS Dartmouth.

Chapter 2: Linking Real-World Data to SAS Event Stream Processing Through Connectors and Adapters

By Vince Deters

Introduction

Event stream processing (ESP) engines that process a continuous flow of real-world data need a pipeline to the sources or sinks that are out in the world. The most common pipeline is a connector or an adapter. Connectors and adapters interface directly with a variety of message brokers, databases, file systems, industrial protocols, and other devices. They have two basic functions: 1) establish connectivity to the outside source or sink, and 2) transform sent or received data to and from discrete events.

For any single instance of a connector or adapter, events must conform to the fixed schema of the window that the instance is configured to write or read. For transmission efficiency, these events are bundled into *event blocks*, which are the basic units of transmission into and out of an ESP window.

Connectors are instantiated in the same process space as the ESP engine and are enabled within the ESP engine's XML-based configuration (Figure 2.1).

Figure 2.1: Connectors

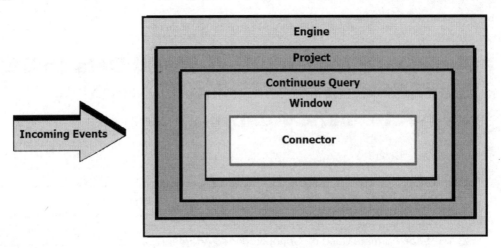

Adapters are stand-alone executables, usually built from corresponding connector classes, and they are started and stopped independently of any ESP engine process. Adapters use the TCP socket-based publish/subscribe API to send and receive event blocks to and from an event stream processing engine (Figure 2.2).

Figure 2.2: Adapters

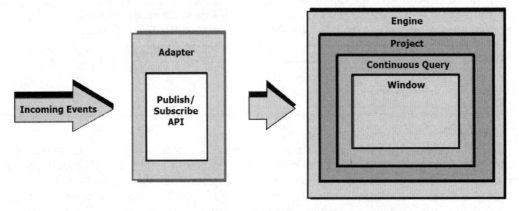

A single instance of a connector or adapter reads from (a subscriber) or writes to (a publisher) a single window in an event stream processing model. Also, a single instance of a connector or adapter is configured as a publisher or subscriber, but not both. Because connectors always run in process with the engine, they are written exclusively in C++, as the engine is.

Adapters that are based on connectors written in C++ simply run an instance of the connector class. There is another set of adapters written in other languages, which run their own self-contained code. Either way, adapters run independently of any engine. They are not in any way synchronized with an engine until they establish a socket connection to a publish/subscribe server running in the engine process.

There is, however, a method by which users can run adapters in process with an engine. This can be useful when an adapter has no corresponding connector, but a user wants to run the adapter as a connector in order to take advantage of connector orchestration. In this case you can run the *adapter connector*, which is a specific connector type that runs an instance of any adapter as an independent spawned process. You can include instances of this adapter connector in your connector orchestration configuration, as you would any other connector instance.

Choosing Between a Connector or an Adapter

When would you use a connector instead of an adapter and vice versa? The answer depends on whether you need connectivity inside the engine or need connectivity on a different computer system or another part of the network. Running adapters on different systems from where you run the engine can be useful when you need to reserve CPU cycles on the system where you are running the engine. It might also be useful when you need connectivity to real-world events outside of a corporate firewall. And finally, you might want to run an adapter as a connector in order to take advantage of orchestration. Note that when you run an adapter on a separate machine from the engine, that machine still requires a full installation of SAS Event Stream Processing.

The Role of Third-party Libraries

In the Internet of Things, every "thing" has an API that governs the connectivity to it. The API is usually provided through third-party libraries. SAS Event Stream Processing often requires you to download these libraries and install them on your server or wherever you are running an adapter. As much as possible, SAS Event Stream Processing provides libraries in order to save customers time and trouble. When the product provides libraries, it controls the version shipped, which ensures compatibility with the data source or sink and the associated connector or adapter. Nevertheless, most third-party libraries are upward and downward compatible.

For the cases where a connector or adapter requires you to independently install a third-party library, you should download the most stable current version unless the product documentation specifies otherwise.

Loading Connectors

All connector classes are derived from a base connector class that is included in a separate connector library. The base connector library includes a connector manager that is responsible for dynamically loading connectors during the initialization of the event stream processing engine. Results of this loading process appear in the engine console log. You can see this same loading process in the adapter console log when running a connector-based adapter.

Connectors are loaded dynamically because they might require third-party libraries that users should only have to install if they are actually using that connector in their model or adapter. All connectors that require users to install third-party libraries are by default not loaded.

Users can modify loading behavior by modifying specific entries in the connectors:excluded section of esp-properties.yml. When a connector that requires third-party libraries is enabled for loading and the required libraries are installed, the console log should reflect that the connector was successfully loaded. When the log shows an error while loading the connector, it is likely that the LD_LIBRARY_PATH (on Linux) or the PATH (on Windows) needs to be modified to include the path to the third-party libraries.

Message Formats Used by Connectors and Adapters

SAS Event Stream Processing provides publish/subscribe API support for the following message formats:

- ESP binary event blocks
- ESP CSV formatted events
- Google Protocol Buffers
- JSON messaging
- Apache Avro messaging
- XML messaging

By default, all connectors and adapters support binary event blocks, and most support CSV events. Most also support transparent conversion to and from ESP event blocks, such as database connectors and adapters, and industrial protocol connectors and adapters that have well known and fixed native formats.

For flexibility, the other listed formats are supported to enable transfer to and from generic sources and sinks that do not mandate any specific format, such as message buses. This is controlled by connector and adapter configuration, which is fixed for the lifetime of the connector or adapter. Thus, a Kafka connector or adapter configured for JSON format (for example) exchanges JSON messages exclusively with the message bus for the lifetime of the connector or adapter. This enables a third-party publisher or subscriber to connect to the message bus and exchange messages with an engine without using the SAS Event Stream Processing publish/subscribe API or having to build ESP CSV events manually.

Be aware, however, that for the names of key fields in key/value entries in such messages to be correctly exchanged, they must match ESP window names. There is no explicit key to window field mapping required or supported.

These format conversions to and from ESP event blocks are implemented in individual libraries, such that different connectors and adapters configured to exchange JSON (for example) can dynamically load the JSON format library one time as needed.

Configuring Connectors and Adapters

Connectivity to an outside source or sink varies by connector or adapter type, but generally requires configuration of some form of user credentials, connectivity parameters, and specification of what data to read or write. For message buses, this would be a topic. Topics provide one-to-many communication on the bus. For database connectivity, this would be a table name.

Configuration consists of required and optional parameters, and these parameter sets are different whether you are running as a publisher or subscriber. The entire set of configuration parameters can be shown on the command line by running an adapter with no arguments.

Connectors are configured by including required and any optional parameters in the engine model XML file. Adapter configuration is specified on the adapter command line. However, both connectors and adapters can read some or all of their configuration from a file through the "configfilesection" parameter.

Some configuration parameters are common to every connector and adapter.

For example, every publisher supports a configurable event block size. The default value is 1, but you might want to set it much higher to optimize throughput, especially on adapters where every transmitted event block incurs TCP overhead. However, a higher block size means higher latency, especially for older events in the block. For low throughput environments where you want to get every event into the model as fast as possible, a block size of 1 is fine.

A common subscriber parameter is "snapshot." When enabled, the subscriber sends the complete contents of the subscribed window before sending any new events generated by the window.

Finally, two common adapter parameters are "transport" and "transportconfigfile." This enables an adapter to use a message bus instead of a direct TCP connection to the ESP engine. The engine must then run a corresponding message bus connector. The advantage here is that the message bus producer only has to write messages one time for multiple consumers to have access to it (that is, a fan-out model). See the "Alternative Client Transports for Adapters" section later in this chapter for more details.

Publishers and Subscribers

Publisher connectors and adapters can run continuously or not, depending on the type. For example, file and database publishers publish the entire contents of the configured file or database query and then stop. Other publishers like those that read from a message bus topic run continuously. All publishers support an optional "maxevents" parameter that stops the publisher when the specified number of events have been published into the source window. All adapters also support an optional "restartonerror" parameter to automatically restart the adapter after it has logged a fatal error during data processing.

Subscriber connectors and adapters run continuously until the engine is stopped by the user. There are two exceptions: 1) a fatal error occurs, or 2) the adapter's TCP connection to the ESP engine breaks due to some external network anomaly. Fatal errors are identified by any connector log message logged at "ERROR" level or higher.

Correspondingly, connectors have three possible states of execution: "stopped," "running," or "finished." The "stopped" state means the connector never started, was stopped because the project was stopped, or was stopped due to a fatal error. The other two states are self-explanatory. These states control the execution flow of connector orchestration, discussed in the sections that follow.

Publisher Source Window Schema Requirements

The source window schema requirements vary by publisher connector or adapter. Some require a schema that directly matches the number of fields and data types available in the data source. For example, for a database or Cassandra or Teradata or CAS or LASR publisher, the schema of the source window must match the schema of the table that you are reading a row of data from.

In other cases, the data source has no schema but just produces a stream of discrete values. This is common in industrial protocols like PI and the OPC connectors and adapters. In this case, you generally need to define a 64-bit integer in your source window schema to act as a key field, and then ensure that you define other fields to match the data type of values you want to read.

Building Events in Publishers

All ESP events must contain opcode and flags parameters. When source data does not include an opcode or flags, the default opcode is "insert" and the default flags is "normal." Publishers can be configured to use "upsert" instead of "insert."

Also consider the case where source data does not contain the field identified as a key field in the source window schema. In this case, publishers can be configured with "noautogenfield," and the source window must be configured to autogenerate the key field. This works only when the window has a single key field.

Parsing Events in Subscribers

Since every event contains an opcode, the opcode must somehow be applied when writing to the data sink. For database subscribers, the opcode translates directly to an insert/update/delete operation on a specific row in the target table. For most other subscribers, the opcode and flags are simply copied to corresponding additional fields in the data sink.

Writing Your Own Connector

If the set of connectors provided with the product does not fulfill your needs, you can write you own connector. When you do, the connector class must inherit from base class dfESPconnector.

Connector configuration is maintained in a set of key-value pairs where all keys and values are text strings. A connector can obtain the value of a configuration item at any time by calling getParameter() and passing the key string. An invalid request returns an empty string.

For more information about the dfESPconnector class, open $DFESP_HOME/doc/html/index.html (UNIX deployments) or %DFESP_HOME%\doc\html\index.html (Windows deployments) in a web browser. This provides access to the complete class and method documentation.

A connector can implement a subscriber that receives events generated by a window or a publisher that injects events into a window. However, a single instance of a connector cannot publish and subscribe simultaneously.

A subscriber connector receives events by using a callback method defined in the connector class that is invoked in a thread owned by the engine. A publisher connector typically creates a dedicated thread to read events from the source. It then injects those events into a Source window, leaving the main connector thread for subsequent calls made into the connector.

Orchestrating Connectors

By default, all connectors start automatically when their associated project starts, and they run concurrently. Connector orchestration enables you to define the order in which connectors within a project execute, depending on the state of another connector. This enables you to create self-contained projects that orchestrate all their inputs. Connector orchestration can be useful to load reference data, inject bulk data into a window before injecting streaming data, or with join windows.

You can represent connector orchestration as a directed graph, just as you represent a continuous query. In this case, the nodes of the graph are connector groups, and the edges indicate the order in which groups execute.

Connectors ordinarily are in one of three states: running, stopped, or finished.

- running – this is the default state of a connector when a project starts
- stopped – a connector is not running because, as a result of connector orchestration, it is waiting for another connector to finish
- finished – a connector has published all of its input events

Subscriber connectors and publisher connectors that are able to publish indefinitely (for example, from a message bus) never reach finished state.

For connector execution to be dependent on the state of another connector, both connectors must be defined in different connector groups. Groups can contain multiple connectors, and all dependencies are defined in terms of the group, not the individual connectors.

When you add a connector to a group, you must specify a corresponding connector state as well. This state defines the target state for that connector within the group. When all connectors in a group reach their target state, all other groups dependent on that group are satisfied. When a group becomes satisfied, all connectors within that group enter running state.

Consider the following configuration that consists of four connector groups: G1, G2, G3, and G4. G1 consists of two connectors, each one in different states, while G2, G3, and G4 each consist of a single connector.

```
G1: {<connector_pub_A, FINISHED>, <connector_sub_B, RUNNING>}
G2: {<connector_pub_C, FINISHED>}
```

```
G3: {<connector_pub_D, RUNNING>}
G4: {<connector_sub_E, RUNNING>}
```

Now consider the following orchestration:

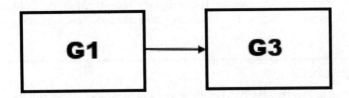

Start the connector in G3 after all the connectors in G1 reach their target states.

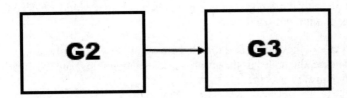

Start the connector in G3 after the connector in G2 reaches its target state. Given the previous orchestration, this means that G3 does not start until the connectors in G1 and in G2 reach their target states.

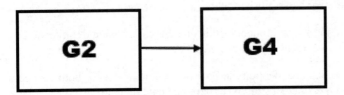

Start the connector in G4 after the connector in G2 reaches its target state.

Because G1 and G2 do not have dependencies on other connector groups, all the connectors in those groups start right away. The configuration results in the following orchestration:

1. When the project is started, `connector_pub_A`, `connector_sub_B`, and `connector_pub_C` start immediately.
2. When `connector_pub_C` finishes, `connector_sub_E` is started.
3. `connector_pub_D` starts only after all conditions for G3 are met, that is, when conditions for G1 and G2 are satisfied. Thus, it starts only when `connector_pub_A` is finished, `connector_sub_B` is running, and `connector_pub_C` is finished.

Alternative Client Transports for Adapters

Adapters can publish or subscribe using a Kafka cluster, Rabbit MQ broker, Tervela Data Fabric, or Solace fabric instead of a direct TCP/IP connection to the ESP server. This is controlled by the "transport" configuration parameter mentioned above. When this parameter is configured, the connectivity parameters required to access the message bus are configured in a file specified by the "transportconfigfile" parameter.

When these parameters are correctly configured, the adapter simply loads the corresponding Kafka, Rabbit MQ, or Solace client transport library at start-up, instead of the ESP native publish/subscribe library. These libraries are located in the "plugins" directory of your ESP distribution and are distinguished from connector plug-in libraries by the "ppi" designation in the library name (publish/subscribe plug-in), versus the "cpi" designation (connector plug-in). These libraries implement the standard publish/subscribe API, so the rest of the adapter and connector code does not know or care that it is using an alternate transport type.

For Java adapters, the act of loading an alternate transport library is achieved by placing the JAR file that implements the configured transport in front of the default native publish/subscribe transport JAR file in the Java classpath. You can see this logic in the Java adapter start-up scripts for linux and windows in your ESP distribution "bin" directory.

The third-party libraries used by these client transport libraries are the same ones used by the corresponding Kafka, Rabbit MQ, and Solace connectors to communicate with the message bus.

These client transports might also load one of the message-format libraries discussed above, if configured to do so in the config file specified by the "transportconfigfile" parameter.

Connectors and Adapters Available with SAS Event Stream Processing

Tables 2.1 through 2.6 group connectors and adapters by usage.

Table 2.1: Connectors and Adapters That Extend SAS Functionality

Name	Description	Connector Available?	Adapter Available?
Adapter Connector	Functions as a wrapper around an adapter, enabling the ESP server to process the adapter as a connector.	Yes	No
SAS Cloud Analytic Services	Enables you to stream data to and from SAS Cloud Analytic Services (CAS). The CAS server is suitable for both on-premises and cloud deployments. It provides the run-time environment for data management and analytics.	No	Yes
Event Stream Processor	Enables you to subscribe to a window and publish what it passes into another Source window.	No	Yes

Name	Description	Connector Available?	Adapter Available?
SAS LASR Analytic Server	Supports publish and subscribe operations on the SAS LASR Analytic Server.	No	Yes
Nurego	Enables you to stream data from Nurego, a cloud-based analytics solution for subscription businesses. The Nurego connector is intended for use with a metering window.	Yes	No
Project Publish	Enables you to subscribe to event blocks that are produced by a window from a different project within the event stream processing model.	Yes	No
Timer	Generates and publishes trigger events at regular intervals.	Yes	Yes

Table 2.2: Connectors and Adapters Used with Audio or Video Devices

Name	Description	Connector Available?	Adapter Available?
Pylon	Communicates with a Basler GigE or USB camera to continuously publish captured frames into a Source window. For GigE cameras, there must be a known, fixed IP address, and the attached Ethernet network cable must provide power using Power-over-Ethernet.	Yes	Yes
UVC	Enables you to publish photos taken by a V4L2–compatible camera to an event stream processing model.	Yes	Yes

Table 2.3: Connectors and Adapters Used with Social Media

Name	Description	Connector Available?	Adapter Available?
BoardReader	Enables you to stream data from a feed aggregator. The site boardreader.com enables you to search message boards, websites, blogs, and other social media.	No	Yes
Twitter	Consumes Twitter streams and injects event blocks into Source windows of an engine.	No	Yes
Twitter Gnip	Consumes data from a Twitter Gnip firehose stream. The adapter is a Python script that invokes the SAS Event Stream Processing C publish/subscribe and JSON libraries.	No	Yes

Table 2.4: Connectors and Adapters Used with Files and Databases

Name	Description	Connector Available?	Adapter Available?
Cassandra	Enables you to stream data to and from Apache Cassandra databases.	No	Yes
Database	Supports publish and subscribe operations to a large number of databases. The DataDirect ODBC drivers are certified for the following databases: Oracle MySQL IBM DB2 Greenplum Hive PostgreSQL SAP Sybase ASE Teradata Microsoft SQL Server IBM Informix Sybase IQ	Yes	Yes
File and Socket	Supports both publish and subscribe operations on files or socket connections that stream a large number of data types.	Yes	Yes
HDAT Reader	SAS software provides a file format for Hadoop called SASHDAT, which makes files available for load to SAS LASR Analytic Server. The HDAT reader adapter converts each row in a SASHDAT file into an ESP event and injects event blocks into a Source window of an engine.	No	Yes
HDFS	Supports publish and subscribe operations to a Hadoop Distributed File System.	No	Yes
SAS Data Set	Connects to a SAS Workspace Server in order to stream data to and from a SAS data set.	No	Yes
Teradata	Enables you to stream data to and from a Teradata server.	Yes	Yes
Teradata Listener	Enables you to stream data from a Teradata Listener, which ingests and distributes extremely fast moving data streams throughout an analytical ecosystem,	Yes	Yes

Table 2.5: Connectors and Adapters Used with Message Queues and Protocols

Name	Description	Connector Available?	Adapter Available?
MQTT	Enables you to stream data through MQ Telemetry Transport (MQTT), a simple and lightweight publish/subscribe messaging protocol designed for constrained devices and low-bandwidth, high-latency, or unreliable networks.	Yes	Yes
Java Message Service (JMS)	Bundles the JMS publisher and subscriber clients. JMS is a Java Message Oriented Middleware (MOM) API to send messages between two or more clients.	No	Yes
Kafka	Enables you to stream data to and from an open-source streaming platform developed by the Apache Software Foundation. The platform is engineered to publish and subscribe to streams of records, similar to a message queue or an enterprise messaging system.	Yes	Yes
Kinesis	Enables you to use Amazon Kinesis Data Streams to collect and process streams of data records in real time.	Yes	Yes
RabbitMQ	Enables you to stream data to and from RabbitMQ, a lightweight open-source message broker.	Yes	Yes
REST	Supports subscribe operations to generate HTTP POST requests to a configured REST service.	No	Yes
SMTP	Enables you to stream data through the Simple Mail Transfer Protocol, an internet standard for electronic mail (email) transmission.	Yes	Yes
Solace	Enables you to stream data to and from Solace, a message broker to establish event-driven interactions between applications and microservices.	Yes	Yes
Tibco Rendezvous	Enables you to stream data to and from Tibco RV, a software product that provides a message bus for enterprise application integration (EAI).	Yes	Yes
URL	Consumes data into a project through a URL-based connection.	Yes	No
WebSocket	Enables you to read data over a WebSocket-based connection and publish the data into a project.	Yes	No
IBM WebSphere	Supports the IBM WebSphere Message Queue Interface for publish and subscribe operations.	Yes	Yes

Table 2.6: Connectors and Adapters for Industrial Use

Name	Description	Connector Available?	Adapter Available?
BACNet	Enables you to stream data over a communications protocol for Building Automation and Control (BAC) networks. The protocol is based on the ISO 16484-5 standard protocol.	Yes	Yes
Modbus	Enables you to stream data to and from Modbus, a serial communications protocol commonly used to connect industrial electronic devices.	Yes	Yes
OPC-DA	Enables you to publish through OPC Data Access (OPC-DA). OPC-DA is machine-to-machine protocol that is a precursor to OPC Unified Architecture (OPC-UA).	No	Yes
OPC-UA	Enables you to stream data through OPC Unified Architecture (OPC-UA), a machine-to-machine communication protocol for industrial automation. OPC-UA is used for communication with industrial equipment and systems for data collection and control.	Yes	Yes
PI	Enables you to stream data to and from a PI Asset Framework server. This server is a repository for asset-centric models, hierarchies, objects, and equipment.	Yes	Yes
Sniffer	Enables you to stream data from a packet sniffer. A sniffer is a software program or hardware device that intercepts and logs traffic that passes over a digital network or part of a network.	Yes	Yes

Example: Using a File and Socket Connector and a WebSocket Connector

The following example shows how to use a WebSocket connector to get events from one ESP server and publish them into another. You start the ESP servers on separate UNIX command lines or start them on the same command line, running commands in the background. On another command line, you start the ESP client to publish events from one ESP server to the other.

The WebSocket Protocol enables two-way communication between a client that runs untrusted code in a controlled environment and a remote host that accepts the communication that results from the code. You can read data over a WebSocket-based connection and publish data into SAS Event Stream Processing using the WebSocket connector. The connector supports XML and JSON over the WebSocket connection. It provides a fully functional API that enables you to parse and transform data into streaming events. The WebSocket connector uses event loops to parse, transform, and publish data. For XML and JSON, the connector reads the WebSocket data until it can build an object of the appropriate type. It then uses functions to process the object and publish events.

You can obtain the following example, including the supporting input files, from a download website for SAS Event Stream Processing customers.

The configuration file for the WebSocket connector is shown in Figure 2.3.

Figure 2.3: Configuration File for the WebSocket Connector

```
<websocket-connector>
    <generate>0</generate>
    <event-loop-json name='events'>
        <use-json>#_content</use-json>
        <json>$.events</json>
        <function-context>
            <functions>
                <function name='broker'>json(#_context,'event.broker')</function>
                <function name='brokerAddress'>json(#_context,'event.brokerAddress')</function>
                <function name='brokerEmail'>json(#_context,'event.brokerEmail')</function>
                <function name='brokerName'>json(#_context,'event.brokerName')</function>
                <function name='brokerage'>json(#_context,'event.brokerage')</function>
                <function name='buyer'>json(#_context,'event.buyer')</function>
                <function name='buysellflg'>json(#_context,'event.buysellflg')</function>
                <function name='closeSeconds'>json(#_context,'event.closeSeconds')</function>
                <function name='closeTimeGMT'>json(#_context,'event.closeTimeGMT')</function>
                <function name='currency'>json(#_context,'event.currency')</function>
                <function name='date'>json(#_context,'event.date')</function>
                <function name='id'>json(#_context,'event.id')</function>
                <function name='msecs'>json(#_context,'event.msecs')</function>
                <function name='openSeconds'>json(#_context,'event.openSeconds')</function>
                <function name='openTimeGMT'>json(#_context,'event.openTimeGMT')</function>
                <function name='price'>json(#_context,'event.price')</function>
                <function name='quant'>json(#_context,'event.quant')</function>
                <function name='seller'>json(#_context,'event.seller')</function>
                <function name='symbol'>json(#_context,'event.symbol')</function>
                <function name='time'>json(#_context,'event.time')</function>
                <function name='timeAfterOpen'>json(#_context,'event.timeAfterOpen')</function>
                <function name='timeTillClose'>json(#_context,'event.timeTillClose')</function>
                <function name='tradeSeconds'>json(#_context,'event.tradeSeconds')</function>
                <function name='venue'>json(#_context,'event.venue')</function>
            </functions>
        </function-context>
    </event-loop-json>
</websocket-connector>
```

Notice that the file includes an event loop (within the `<event-loop-json>` element) that references JSON in order to generate events. Within the `<function-context>` element are entities to run JSON on event data and generate values for output events. The `<function>` elements specify output event fields to generate.

The model in Figure 2.4 contains multiple Source windows to read in broker, venue, trade, and restricted trade data. Each Source window uses a file and socket connector to receive input event streams. The model uses a Counter window to maintain a time-based count of events, and uses a combination of Join, Functional, and Filter windows to transform the event streams.

Figure 2.4: Input Model

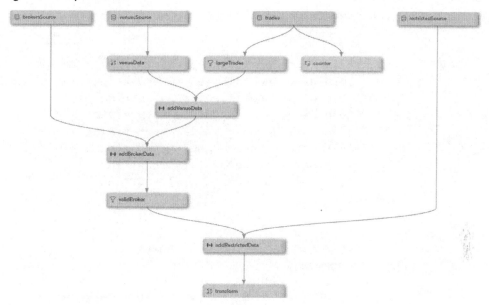

The XML that specifies the windows of this model is presented in a series of figures that follow. The model obtains its input events from various CSV files specified in the XML code for the file and socket connector.

There are four Source windows: one to input broker events (Figure 2.5), a second to input venue events (Figure 2.6), a third to input trade events (Figure 2.7), and a fourth to input restricted trade events (Figure 2.8).

Figure 2.5: Source Window for Broker Events

```
<window-source name='brokersSource' insert-only='true'>
    <schema>
        <fields>
            <field name='broker' type='int32' key='true'/>
            <field name='brokerName' type='string'/>
            <field name='brokerage' type='string'/>
            <field name='brokerAddress' type='string'/>
            <field name='brokerEmail' type='string'/>
            <field name='brokerPhone' type='string'/>
            <field name='brokerSms' type='string'/>
            <field name='brokerMms' type='string'/>
        </fields>
    </schema>
    <connectors>
        <connector class='fs' name='connector'>
            <properties>
                <property name='type'>pub</property>
                <property name='fstype'>csv</property>
                <property name='fsname'>brokers.csv</property>
            </properties>
        </connector>
    </connectors>
</window-source>
```

Figure 2.6: Source Window for Venue Data

```
<window-source name='venuesSource' insert-only='true'>
    <schema-string> venue*:int32,openTimeGMT:string,closeTimeGMT:string
    </schema-string>
    <connectors>
        <connector class='fs' name='connector'>
            <properties>
                <property name='type'>pub</property>
                <property name='fstype'>csv</property>
                <property name='fsname'>venues.csv</property>
            </properties>
        </connector>
    </connectors>
</window-source>
```

Figure 2.7: Source Window for Trades Events

```
<window-source name='trades' insert-only='true'>
    <schema-string> id*:int64,symbol:string,currency:int32,time:int64,
        msecs:int32,price:double, quant:int32,venue:int32,broker:int32,
        buyer:int32,seller:int32,buysellflg:int32
    </schema-string>
</window-source>
```

Figure 2.8: Source Window for Restricted Trades Events

```
<window-source name='restrictedSource' insert-only='true'>
    <schema-string> symbol*:string,venue*:int32,restricted:int32
    </schema-string>
    <connectors>
        <connector class='fs' name='connector'>
            <properties>
                <property name='type'>pub</property>
                <property name='fstype'>csv</property>
                <property name='fsname'>restricted.csv</property>
            </properties>
        </connector>
    </connectors>
</window-source>
```

A Functional window named venueData performs time calculations on the streaming venue events from venueSource (Figure 2.9).

Figure 2.9: Functional Window Processing Venue Events

```
<window-functional name='venueData'>
    <schema-string>
    venue*:int32,openTimeGMT:string,closeTimeGMT:string,openSeconds:int64,closeSeconds:int64
    </schema-string>
    <function-context>
        <functions>
            <function name='openSeconds'>timeSecondOfDay(timeGmtToLocal(
            timeParse($openTimeGMT,'%H:%M:%S')))
            </function>
            <function name='closeSeconds'>timeSecondOfDay(timeGmtToLocal(
            timeParse($closeTimeGMT,'%H:%M:%S')))
            </function>
        </functions>
    </function-context>
</window-functional>
```

Trade events stream into a Filter window named largeTrades (Figure 2.10). It permits trades greater than or equal to 1000 shares to pass through.

Figure 2.10: Filter Window for Large Trades

```
<window-filter name='largeTrades'>
    <expression><![CDATA[quant>=1000]]></expression>
</window-filter>
```

The output of the Filter window and the Functional window streams into a Join window, where selected fields of the streaming events are combined (Figure 2.11).

Figure 2.11: Join Window for Trade Events and Venue Events

```
<window-join name='addVenueData'>
    <join type='leftouter' no-regenerates='true'>
        <conditions>
            <fields left='venue' right='venue'/>
        </conditions>
    </join>
    <output>
        <field-expr name='broker' type='int32'>l_broker</field-expr>
        <field-expr name='buyer' type='int32'>l_buyer</field-expr>
        <field-expr name='buysellflg' type='int32'>l_buysellflg</field-expr>
        <field-expr name='currency' type='int32'>l_currency</field-expr>
        <field-expr name='msecs' type='int32'>l_msecs</field-expr>
        <field-expr name='price' type='double'>l_price</field-expr>
        <field-expr name='quant' type='int32'>l_quant</field-expr>
        <field-expr name='seller' type='int32'>l_seller</field-expr>
        <field-expr name='symbol' type='string'>l_symbol</field-expr>
        <field-expr name='time' type='int64'>l_time</field-expr>
        <field-expr name='venue' type='int32'>l_venue</field-expr>
        <field-expr name='closeSeconds' type='int64'>r_closeSeconds</field-expr>
        <field-expr name='closeTimeGMT' type='string'>r_closeTimeGMT</field-expr>
        <field-expr name='openSeconds' type='int64'>r_openSeconds</field-expr>
        <field-expr name='openTimeGMT' type='string'>r_openTimeGMT</field-expr>
    </output>
</window-join>
```

Trade events also stream into a Counter window (Figure 2.12).

Figure 2.12: Counter Window

```
<window-counter name='counter' count-interval='2 seconds' clear-interval='30 seconds'/>
```

The output from brokersSource and addVenueData streams into another Join window named addBrokerData. Selected fields from the input events are combined (Figure 2.13).

Figure 2.13: Join Window Combining Broker Events and Processed Venue Events

```
<window-join name='addBrokerData'>
    <join type='leftouter' no-regenerates='true'>
        <conditions>
            <fields left='broker' right='broker'/>
        </conditions>
    </join>
    <output>
        <field-expr name='broker' type='int32'>l_broker</field-expr>
        <field-expr name='buyer' type='int32'>l_buyer</field-expr>
        <field-expr name='buysellflg' type='int32'>l_buysellflg</field-expr>
        <field-expr name='closeSeconds' type='int64'>l_closeSeconds</field-expr>
        <field-expr name='closeTimeGMT' type='string'>l_closeTimeGMT</field-expr>
        <field-expr name='currency' type='int32'>l_currency</field-expr>
        <field-expr name='msecs' type='int32'>l_msecs</field-expr>
        <field-expr name='openSeconds' type='int64'>l_openSeconds</field-expr>
        <field-expr name='openTimeGMT' type='string'>l_openTimeGMT</field-expr>
        <field-expr name='price' type='double'>l_price</field-expr>
        <field-expr name='quant' type='int32'>l_quant</field-expr>
        <field-expr name='seller' type='int32'>l_seller</field-expr>
        <field-expr name='symbol' type='string'>l_symbol</field-expr>
        <field-expr name='time' type='int64'>l_time</field-expr>
        <field-expr name='venue' type='int32'>l_venue</field-expr>
        <field-expr name='brokerAddress' type='string'>r_brokerAddress</field-expr>
        <field-expr name='brokerEmail' type='string'>r_brokerEmail</field-expr>
        <field-expr name='brokerMms' type='string'>r_brokerMms</field-expr>
        <field-expr name='brokerName' type='string'>r_brokerName</field-expr>
        <field-expr name='brokerPhone' type='string'>r_brokerPhone</field-expr>
        <field-expr name='brokerSms' type='string'>r_brokerSms</field-expr>
        <field-expr name='brokerage' type='string'>r_brokerage</field-expr>
    </output>
</window-join>
```

The output from this Join window flows into a Filter window named validBroker. This ensures the exclusion of invalid broker data (Figure 2.14).

Figure 2.14: Filter Window for Broker Data

```
<window-filter name='validBroker'>
    <expression>isnull(brokerName)==false</expression>
</window-filter>
```

Events from this Filter window are combined with events from the Source window for restricted trades in another Join window (Figure 2.15).

Figure 2.15: Join Window Combining Broker Events with Restricted Trades

```
<window-join name='addRestrictedData'>
    <join type='leftouter' no-regenerates='true'>
        <conditions>
            <fields left='symbol' right='symbol'/>
            <fields left='venue' right='venue'/>
        </conditions>
    </join>
    <output>
        <field-expr name='broker' type='int32'>l_broker</field-expr>
        <field-expr name='brokerAddress' type='string'>l_brokerAddress</field-expr>
        <field-expr name='brokerEmail' type='string'>l_brokerEmail</field-expr>
        <field-expr name='brokerMms' type='string'>l_brokerMms</field-expr>
        <field-expr name='brokerName' type='string'>l_brokerName</field-expr>
        <field-expr name='brokerPhone' type='string'>l_brokerPhone</field-expr>
        <field-expr name='brokerSms' type='string'>l_brokerSms</field-expr>
        <field-expr name='brokerage' type='string'>l_brokerage</field-expr>
        <field-expr name='buyer' type='int32'>l_buyer</field-expr>
        <field-expr name='buysellflg' type='int32'>l_buysellflg</field-expr>
        <field-expr name='closeSeconds' type='int64'>l_closeSeconds</field-expr>
        <field-expr name='closeTimeGMT' type='string'>l_closeTimeGMT</field-expr>
        <field-expr name='currency' type='int32'>l_currency</field-expr>
        <field-expr name='msecs' type='int32'>l_msecs</field-expr>
        <field-expr name='openSeconds' type='int64'>l_openSeconds</field-expr>
        <field-expr name='openTimeGMT' type='string'>l_openTimeGMT</field-expr>
        <field-expr name='price' type='double'>l_price</field-expr>
        <field-expr name='quant' type='int32'>l_quant</field-expr>
        <field-expr name='seller' type='int32'>l_seller</field-expr>
        <field-expr name='symbol' type='string'>l_symbol</field-expr>
        <field-expr name='time' type='int64'>l_time</field-expr>
        <field-expr name='venue' type='int32'>l_venue</field-expr>
        <field-expr name='restricted' type='int32'>r_restricted</field-expr>
    </output>
</window-join>
```

Events from that Join window stream into another Functional window (Figure 2.16).

Figure 2.16: Functional Window Processing Joined Events

```
<window-functional name='transform'>
    <schema-string>
    id*:int64,broker:int32,brokerAddress:string,brokerEmail:string,brokerMms:string,
    brokerName:string,brokerPhone:string,brokerSms:string,brokerage:string,buyer:int32,
    buysellflg:int32,closeSeconds:int64,closeTimeGMT:string,currency:int32,msecs:int32,
    openSeconds:int64,openTimeGMT:string,price:double,quant:int32,seller:int32,symbol:string,
    time:int64,venue:int32,restricted:int32,date:date,tradeSeconds:int32,timeAfterOpen:int32,
    timeTillClose:int32
    </schema-string>
    <function-context>
        <functions>
            <function name='date'>$time</function>
            <function name='tradeSeconds'>timeSecondOfDay($time)</function>
            <function name='timeAfterOpen'>diff($tradeSeconds,$openSeconds)</function>
            <function name='timeTillClose'>diff($closeSeconds,$tradeSeconds)</function>
        </functions>
    </function-context>
</window-functional>
```

The model in Figure 2.17 contains a single Source window using a WebSocket connector.

Figure 2.17: Model with WebSocket Connector

The XML specifying the entire project is presented in Figures 2.18. It is contained in a file named model2.xml. The model obtains its input events from the first model through a WebSocket connector.

Figure 2.18: XML for Model with WebSocket Connector

```
<project name='p' index='pi_EMPTY' pubsub='auto' threads='4'>
    <contqueries>
        <contquery name='cq' trace='trades'>
            <windows>
                <window-source name='trades' insert-only='true'>
                    <schema-string> id*:int64,broker:int32,brokerAddress:string,brokerEmail:string,
                    brokerMms:string,brokerName:string,brokerPhone:string,brokerSms:string,brokerage:string,
                    buyer:int32,buysellflg:int32,closeSeconds:int64,closeTimeGMT:string,currency:int32,
                    msecs:int32,openSeconds:int64,openTimeGMT:string,price:double,quant:int32,seller:int32,
                    symbol:string,time:int64,venue:int32,restricted:int32,date:string,tradeSeconds:int32,
                    timeAfterOpen:int32,timeTillClose:int32
                    </schema-string>
                    <connectors>
                        <connector class='websocket'>
                            <properties>
                                <property name='type'>pub</property>
                                <property name='url'><![CDATA[ws://localhost:@port@/SASESP/subscribers/p/cq/
                                transform?mode=streaming&format=xml&pagesize=0]]></property>
                                <property name='contentType'>xml</property>
                                <property name='configUrl'>file://config.xml</property>
                            </properties>
                        </connector>
                    </connectors>
                </window-source>
            </windows>
        </contquery>
    </contqueries>
</project>
```

After both models are running, you can use the ESP client to inject 10000 events into the first model. These events are specified in a file named trades.csv. After you run the following command line, you see streaming events from the second model on the console.

Example Code 2.1

```
$DFESP_HOME/bin/dfesp_xml_client -url
"http://localhost:54321/SASESP/windows/p/cq/trades/state?value=injecte
d&eventUrl=file://trades.csv" -put


<event opcode="insert" window="p/cq/trades">
  <value name="broker">1012001</value>
  <value name="brokerAddress">SAS Campus Drive Cary NC 27513</value>
  <value name="brokerEmail">919-123-4567</value>
  <value name="brokerName">Steve</value>
  <value name="brokerage">ESP</value>
```

```
    <value name="buyer">872415284</value>
    <value name="buysellflg">1</value>
    <value name="closeSeconds">54000</value>
    <value name="closeTimeGMT">20:00:00</value>
    <value name="currency">87236</value>
    <value name="date">1280928700</value>
    <value name="id">7237809</value>
    <value name="msecs">1471</value>
    <value name="openSeconds">30600</value>
    <value name="openTimeGMT">13:30:00</value>
    <value name="price">19.188800</value>
    <value name="quant">2000</value>
    <value name="seller">8377283</value>
    <value name="symbol">GLW</value>
    <value name="time">1280928700</value>
    <value name="timeAfterOpen">3700</value>
    <value name="timeTillClose">19700</value>
    <value name="tradeSeconds">34300</value>
    <value name="venue">55999</value>
</event>
```

Notice that the structure of output events matches the function context that is specified in the configuration file in Figure 2.3.

Conclusion

This chapter provides a high-level introduction into the connectivity enabled by connectors and adapters. It only scratches the surface of their complexity and versatility, which is described in detail in the product documentation at https://support.sas.com/en/software/event-stream-processing-support.html#documentation.

The "things" of the Internet of Things use connectors and adapters to stream real-time events into SAS Event Stream Processing, providing the raw material for actionable intelligence at the edge. That intelligence is applied to streaming data by the analytics that are covered in the next chapter.

About the Contributor

As a software developer for ESP on the IOT team at SAS, **Vince Deters** focuses on connectivity features for ESP, including connectors, adapters, and pubsub APIs. Vince is a Duke graduate, has over 30 years of development experience, and is an inventor on numerous patents at SAS and Cisco, where he worked prior to coming to SAS.

Chapter 3: Applying Analytics to Streaming Data

By Gül Ege, Bennett McAuley, and Michael Harvey

Introduction

> A significant predictor of an organization's ability to deliver value from the
> IoT across an enterprise is the heavy use of AI. The true value of IoT data is
> only realized when combined with AI and analytics.
>
> (IDC Study 2019)

Every second of every day, tremendous amounts of data are generated by things in the world: computing devices, industrial machines, uniquely identified people, and so on. The Internet of Things (IoT) creates a connected world with these things. The streaming data generated from these things can create previously unobtainable business value. Proper analysis of the data can transform work in multiple domains, including industry, business, environment, human health, and wildlife management.

Streaming data gathered from the connected world has distinct characteristics: it is very high volume and high frequency, extremely noisy, high dimensional, often missing many observations, and sometimes riddled with redundancies. The data can be numeric, binary, text, audio, video, or image. Furthermore, the statistical distributions of the data are not always known and cannot be assumed. Harnessing the potential of this data into actionable insights at the edge, in the fog, and in the cloud requires a creative application of advanced analytic techniques.

Recent innovations in this area have focused on the ability to create actionable insights on streaming data in real-time. The new capabilities are chosen and implemented with the end goal of supporting intelligent action *as data is streaming in*. The advanced analytics offered by SAS Event Stream Processing (ESP) address multiple use cases across many advanced analytical fields to support real-time analysis of streaming data. They take a multi-phased analytical approach to insert the streaming data into the analytics life cycle, augmenting more traditional analytics techniques that are applied to static data.

This chapter provides a review of innovations in advanced analytical models that are trained on static data and scored on streaming data, along with those that are directly applied to streaming data. This creates a multi-phase analytics life cycle.

The Multi-Phase Analytics Life Cycle

To initiate the multi-phase analytics life cycle, the first step is to ask the following questions about the data itself:

- How do I remove noise from the data?
- How much of this data should I store?
- Which part of this data can create insights for me?
- Which features should I create from streaming data before storing the streaming data permanently?

These questions must be answered in terms of the nature of the data and the value-adding insights and actions to be created. Generic answers that cover all use cases simply do not exist. There is also no point in using *all* the data generated by the sensors for analysis. The typical approach is to use a sample of the original data in order to begin addressing these questions. You must determine which aspects of the data can provide predictive knowledge before you put the filters, transformations, and features into production in SAS Event Stream Processing. Dimension reduction and denoising are pivotal to the overall success of any IoT project.

After satisfying inquiries about the data, the next step is *preprocessing* – manipulating and cleaning raw data that might be unsuitable for a meaningful analysis. This step is the most critical and can be the most time-consuming. Fortunately, it needs to be conducted only once, provided that the structure of the source data does not substantially change. SAS enables you to focus on how much of the original data to store and how to denoise IoT data.

In a typical analytics life cycle, data is captured, stored, scored, and analyzed. Models are trained and developed. Results are made available to applications and reporting tools that deliver alerts and support decisions. This approach has an inherent time lag: the lag between when the analysis detects a problem and when an action is taken (Figure 3.1).

Figure 3.1: Typical Analytics Life Cycle

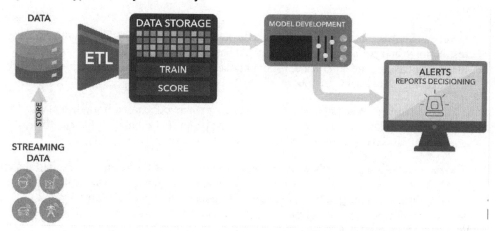

Multi-phased analytics enables you to eliminate the time lag. The alerts can be applied almost immediately because the analysis takes place physically close to the device that generates the data (Figure 3.2).

Figure 3.2: Multi-Phased Analytics Life Cycle

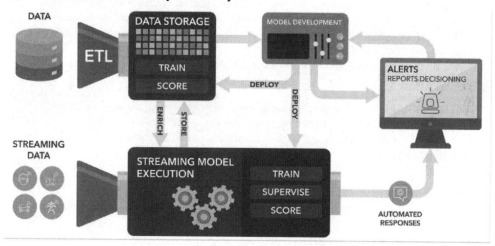

For example, suppose you monitor assets in an industrial plant. The data about a single asset could include temperature, vibration, and RPMs so that you can identify anomalies in its operation. In this case, the first phase of analytics would apply a model to incoming data from the asset in order to detect errors in its performance. You score data coming from the asset by running the model in SAS Event Stream Processing on an IoT gateway installed on the asset. You can use SAS Event Stream Processing on that gateway to retrain that model to better predict the chance of failure. Now suppose you want to monitor multiple assets across an entire factory floor. The second phase of analytics is when SAS Event Stream Processing is running on premises or in the cloud and you use it to analyze data from the entire set of assets. You do cross-entity analytics by aggregating the data from all the IoT gateways installed on assets. Another phase of analytics occurs when you analyze stored, historical data.

Here, an important distinction must be made between data at rest and streaming data. Data at rest is stored and periodically updated. It is most often low volume and low frequency. Streaming data, on the other hand, is continuous, very high volume and frequency. Much of the streaming data might never be stored.

SAS Event Stream Processing enables you to take what you learned from data at rest, train new models, update existing ones, and bring the new or revised models back to the IoT gateway installed on the asset. And this is all done dynamically.

The next sections detail methods for creating value for business based on meaningful alerts. These alerts arise from catching anomalies, finding patterns, and tracking and identifying objects. Results enable you to make predictions about future events. You can make recommendations supported by meaningful analytical techniques capable of scaling and matching the speed of streaming data.

Online and Offline Models

The analytical capabilities of SAS Event Stream Processing are delivered as online and offline models. Online models can be trained and scored within SAS Event Stream Processing as they are installed with the product. Offline models are those generated by other SAS products such as SAS Visual Data Mining and Machine Learning with data at rest. They can be imported into SAS Event Stream Processing to score streaming data. The use of an online versus an offline model depends on the problem, data, and desired algorithm performance.

Figure 3.3 displays a high-level classification of the analytical capabilities in SAS ESP, separated by the most prevalent categories. These capabilities come from many branches of advanced analytics: statistics, time series analysis, machine learning, text mining, deep learning, and so on.

Figure 3.3: Summary of the Analytical Capabilities of SAS Event Stream Processing

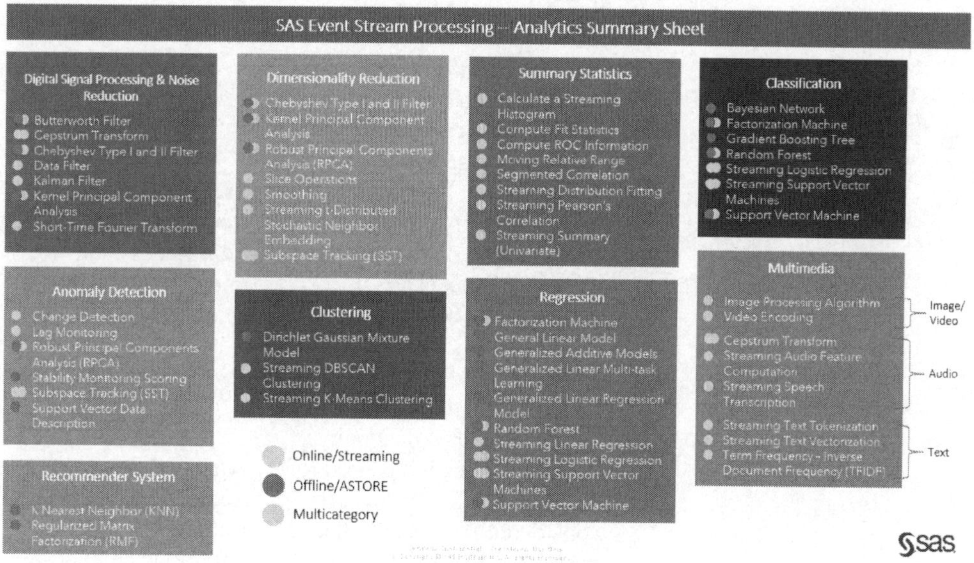

Table 3.1 provides more detail about the categories with some examples of domains that they are commonly associated with and used in.

Table 3.1: Description of the Analytical Capabilities of SAS Event Stream Processing

Category	Description	Use Case Examples
Digital Signal Processing and Noise Reduction	Digital signal processing is the process of optimizing the quality of signals and communications. Data that can be represented as a signal wave (audio, temperature, heartbeat, and so on) is manipulated and denoised using smoothing and approximation algorithms.	Sound compression and quality Speech recognition Mobile communication Medical imaging Biosensors
Dimensionality Reduction	Dimensionality reduction techniques are applied in the preliminary stage of analysis to simplify and enhance the more insightful features of data. This makes the data more useable to work with and optimizes calculation performance (That is algorithm time complexity and storage space).	Facial recognition Radar signal quality Network synthesis Neuroscience

Category	Description	Use Case Examples
Summary Statistics	Summary statistics are measures that provide meaningful information about the data being analyzed beyond what can be inferred by manual inspection.	--
Classification	Classification algorithms are used to identify what group (class) a data point belongs to. Generally, data is classified using a set of attributes that best distinguish and define a particular class. The goal is to maximize the number of correctly labeled data points.	Credit scoring Document classification Computer vision Image segmentation Search engines
Anomaly Detection	Anomaly detection is used to identify data points that are considered abnormal to the pattern of the rest of the data. Often these data points are referred to as outliers.	Financial fraud Engine/machine stability Health monitoring Surveillance Vehicle sensors
Clustering	Clustering algorithms are best used for tasks involving grouping data objects into clusters that exhibit the most similarity. The goal is to maximize distinctness between similar objects in a cluster and differences between clusters.	Genetic sequencing Wireless networks Medical imaging Market research Geostatistics
Regression	Regression algorithms are used to examine relationships between variables. Most commonly, this entails estimating the value of a dependent variable based on the values of independent variables that are hypothesized to have an impact on the dependent variable.	Sales forecasting Market price prediction Health analysis Financial and insurance risk Biological systems
Media	Media refers to algorithms applied to unstructured, visual data. This includes processing, parsing, and manipulating videos, images, audio, and text.	Speech transcription Format/file type conversion Text analytics

Category	Description	Use Case Examples
Recommender System	A recommender system attempts to predict the rating or preference that someone would give an item such as a book, article, or product based on previous ratings or other items.	Customer reviews Netflix recommendations

There are limitless ways to classify the analytical functionalities in ESP. The following list exemplifies a few ways:

- The domain of analytical approaches (time series, frequency, machine learning, deep learning, statistics, and so on)
- The purpose the functionality is to serve (clustering, classification, denoising, dimension reduction, anomaly detection, and so on)
- Where the models are trained and scored (online models versus offline models)

Online Versus Offline Model Deployment

Online models are algorithms packaged with ESP and trained in ESP projects. They use Calculate, Score, and Train windows to implement streaming analytics. When an algorithm is executed with one of these windows, its name, parameter properties, and input/output mappings must be specified.

Calculate windows create real-time, running statistics based on established analytical techniques. The windows receive data events and publish newly transformed score data into output events. Figure 3.4 shows an example of a query calculating Streaming Pearson's Correlation on streaming data. The XML code provides details about the Calculate window in the query.

Figure 3.4: Continuous Query of Streaming Pearson's Correlation

Example Code 3.1

```
<window-calculate name="w_calculate" algorithm="Correlation">
     <schema>
          <fields>
               <field name='id' type='int64' key='true'/>
               <field name='y_c' type='double'/>
               <field name='x_c' type='double'/>
               <field name='corOut' type='double'/>
```

```
                    </fields>
            </schema>
            <parameters>
                    <properties>
                            <property name="windowLength">5</property>
                    </properties>
            </parameters>
            <input-map>
                    <properties>
                            <property name="x">x_c</property>
                            <property name="y">x_c</property>
                    </properties>
            </input-map>
            <output-map>
                    <properties>
                            <property name="corOut">corOut</property>
                    </properties>
            </output-map>
            <connectors>

            </connectors>
</window-calculate>
```

As shown, the Source window receives the data to be analyzed. The Calculate window calculates the correlation between two variables from an incoming data stream and publishes the results in real-time. In the code, the Correlation algorithm is called to execute on the w_calculate window. The specifications for the fields, parameters, input-map, and output-map elements depend on the algorithm. *Field* elements specify a definition or a reference to a column in an event. *Parameter* elements specify the parameter values that govern the algorithm. Streaming Pearson's Correlation has one parameter, windowLength, which determines the number of events used in the calculation. *Input-map* elements specify the variable names that the algorithm expects from incoming events. *Output-map* elements specify the variable names that the algorithm produces and publishes to another window or an output file.

Other online algorithms in ESP, such as k-means clustering, use Train and Score windows in tandem. Train windows receive data events and publish model events to Score windows. Score windows accept model events to make predictions for incoming data events and generate scored data. Figure 3.5 shows an example of a Train and Score query for k-means with its associated XML code.

Figure 3.5: Continuous Query of Streaming K-means Clustering

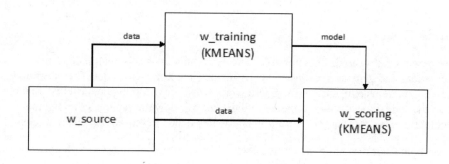

Example Code 3.2

```
<window-train name='w_training' algorithm='KMEANS'>
     <parameters>
          <properties>
               <property name="nClusters">2</property>
               <property name="initSeed">1</property>
               <property name="dampingFactor">0.8</property>
               <property name="fadeOutFactor">0.05</property>
               <property name="disturbFactor">0.01</property>
                <property name="nInit">50</property>
               <property name="velocity">5</property>
               <property name="commitInterval">25</property>
          </properties>
      </parameters>
     <input-map>
          <properties>
               <property name="inputs">
                <![CDATA[x_c,y_c]]>
               </property>
          </properties>
     </input-map>
</window-train>

<window-score name='w_scoring'>
     <schema>
          <fields>
               <field name='id'  type='int64' key='true'/>
               <field name='x_c' type='double'/>
                <field name='y_c' type='double'/>
               <field name='seg' type='int32'/>
               <field name='min_dist' type='double'/>
               <field name='model_id' type='int64'/>
          </fields>
     </schema>
     <models>
          <online algorithm='KMEANS'>
               <input-map>
                <properties>
                 <property name="inputs">
                  <![CDATA[id, x_c, y_c, seg, min_dist,model_id]]>
```

```
                           </property>
                       </properties>
                  ·  </input-map>
               </online>
          </models>
</window-score>
```

The Source window holds the same role. The Train window looks at all incoming observations and generates and periodically updates the k-means model. The generated clustering model is published to the Score window, where incoming data is clustered. Notice that in the code, the algorithm is called in the Train window the same way it is in the Calculate window, but for the Scoring window, it is called in the *models* element tags.

To see more examples of online models and further information, refer to *SAS Event Stream Processing: Using Streaming Analytics* in the SAS documentation.

Offline models that can be processed in ESP follow a different approach to model deployment. Offline models are specified, developed, trained, and stored separately from the ESP server. ESP supports a variety of sources for offline models such as SAS Visual Data Mining and Machine Learning (VDMML), SAS Visual Text Analytics, JMP, and open-source frameworks like TensorFlow. Figure 3.6 depicts how offline models are developed and managed for use with ESP.

Figure 3.6: Offline Analytical Model Deployment and Management

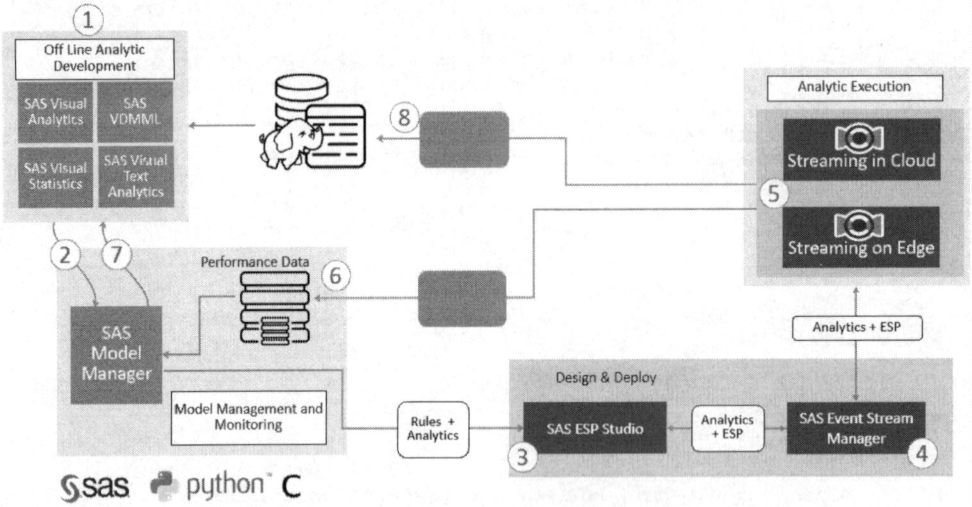

To implement offline models with streaming data in ESP, a Model Reader window is used. Model Reader windows receive request events that include that the location and type of offline model being applied. A Source window streams data into the Model Reader window through a request event. The Model Reader window then streams the offline model into a Score window through a model event. Figure 3.7 provides a visual representation of this process. You can use a separate Source window to stream data to be stored by the model into the Score window through a data event.

Figure 3.7: Continuous Query of an Offline Model Processed in ESP

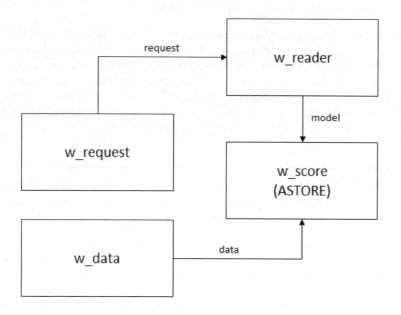

Example Code 3.3

```
<window-model-reader name='w_reader' model-type='astore'/>
```

In the figure, the Model Reader window named w_reader receives requests from the Source window, w_request, and fetches the specified model using the request information. ASTORE is specified as the model-type, meaning that the model is coming from a source outside of the ESP server. Most offline models developed with SAS software are stored in analytic store (ASTORE) files. It is a binary file that contains the model's state after it completes training. Other offline models, such as Recommender Scoring, are stored in combinations of non-binary files.

For more information, see *SAS Event Stream Processing: Online Scoring and Training Using Offline Models* in the SAS documentation.

Potential for Model Application

One of the common, high-level goals across different use cases is to create accurate and timely alerts in real-time. ESP also enables domain knowledge to be expressed as rules that lead to these alerts. Alerts based only on domain rules often have a shortcoming of over-alerting because of noise in the data.

A specific application for alerting is anomaly detection. Referring back to Table 3.1, anomaly detection is often used for detecting fraud, monitoring system/machine stability, and surveillance. Banking, engineering, health care, and security are industry domains that carry significant influence on the well-being of businesses and civilians. For this reason, SAS has focused its efforts on providing methods in advanced analytics for IoT that minimize false positives and improve on lead time to avoid model failure.

Many of the anomaly detection techniques packaged in ESP like Support Vector Data Description (SVDD) and Stability Monitoring are implemented with these goals in mind. Both techniques are offline algorithms that are trained with historic data to capture the "normal state" of a system. The models are scored on streaming data to catch system degradation and projected failure. Subspace tracking (SST), on the other hand, is an online model used for detecting anomalies. The rest of this chapter explains these algorithms in more detail as well as provide a few examples of how they perform.

Stability Monitoring

Stability Monitoring scores events to monitor the stability of a system. It detects anomalous behavior of various signals within event data and, using the stored model, generates forecasts of a target signal.

The analytical approach of Stability Monitoring assumes that within a subset of sensors, events, or both – collectively called *signals* – there exists a "target" signal. The behavior of this target signal can be explained by other, "explanatory" signals. The main idea is that a system is considered stable when there is a robust statistical relationship between the target signal and explanatory signals.

A relevant statistical model is selected, fitted on historical data during stable periods, and saved using the SMCALIB procedure. The stored model is applied to new data that contains both target and explanatory signals to generate a forecast of a target signal. This is where model scoring occurs, and it is performed using the SMSCORE procedure. The forecast is then compared to the actual values of the target signal. Based on a set of user-defined rules, anomalies in the monitored system can be flagged.

Advantages to using Stability Monitoring include the following:

- Multiple models can be specified and evaluated for calibration.
- The algorithm can handle multivariate data and automatically perform variable selection for each model.
- The algorithm uses holdout analysis to evaluate selected model types and perform calibration.

Support Vector Data Description

Support Vector Data Description is a one-class classification technique that can be useful when data about one class is abundant but data about any other class is scarce or missing. You can model one-class data and subsequently use the model for anomaly detection. Data from normal operating conditions are used to create the model, and observations lying outside of the generated boundary are considered anomalies.

In its simplest form, an SVDD model is obtained by building a minimum-radius hypersphere around the one-class training data. A hypersphere is a set of points that are a constant distance away from a defined center point. The hypersphere provides a compact spherical description of the training data that can be used to determine whether a new observation is similar to the training data observations. The distance from any new observation to the hypersphere center is computed and compared with the radius. If the distance is more than the radius, the observation is designated as an outlier. Using kernel functions in SVDD formulation provides a more flexible description of training data. Such a description is nonspherical and conforms to the geometry of the data.

Advantages to using SVDD include the following:

- The algorithm can handle multivariate data.
- The algorithm does not require any assumptions of normality in the data.
- The algorithm has an option to automatically select the value for the kernel bandwidth parameter.
- The algorithm does not require data to be labeled and identifies anomalies that occur outside of normal operating conditions that are used to train the model. Namely, the algorithm can potentially identify anomalies that are rare to occur or might not have occurred in previously processed data.
- The algorithm can be applied in situations where the data is imbalanced between examples of normal operating conditions and deteriorating conditions.

Application of Offline Models on Streaming Data

This section presents a use case for Stability Monitoring and Support Vector Data Description: monitoring engine degradation. The focus is on how sensor data generated from aircraft engines can be used to monitor asset degradation and aid with predictive maintenance efforts. If the multivariate sensor data can be used to determine when an asset is starting to break down, then it can help in optimizing maintenance schedules. It could also help reduce the number of breakdown events that lead to costly repairs, lost productivity, and potential injury. Specifically, this use case shows how engine degradation can be identified for a turbofan engine data set produced by NASA for the 2008 conference on Prognostics and Health Management (Saxena and Goebel 2008).

Stability Monitoring Method

Stability Monitoring uses statistical methods to approach the problem as follows:

- Choose a target variable (sensor) whose behavior should be explained by other sensor variables within the data set.
- Specify and calibrate a variety of statistical models.
- Select the best performing model based on holdout validation.
- Forecast target sensor values for new input values based on the calibrated model.
- Compare the forecasted value versus the actual value with a rule definition to indicate an anomaly dependent on the business context. (For example, the actual value is outside of the 95% confidence band.)

Stability Monitoring Results

The aircraft engine used in this example had 240 cycles associated with it. The first 80 cycles of the engine were used for model training with the next 10 cycles used to perform the holdout validation. The model was then further evaluated on the remaining cycles.

The ratio of fuel flow to static pressure at the high-pressure compressor outlet was chosen as the target sensor with the assumption that as equipment performance begins to deteriorate, it is reflected in changes in the fuel consumption.

Figure 3.8 shows the output from the method on the turbofan data. Each graph represents a different time frame where the blue circles, red circles, and shaded zones represent actual values, anomalies, and the 95% prediction limits, respectively. The top left quadrant of the figure illustrates the holdout zone that was used to calibrate the model.

Figure 3.8: Stability Monitoring Results

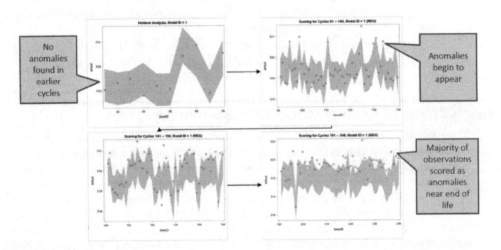

As seen in the figure, all the actual values are within the predicted limits of the area, so no anomalies are present. In the top right of the figure, we begin to see anomalies appear. These anomalies become more prevalent in the subsequent time frame in the lower left quadrant and continue to within the final time frame in the lower right quadrant.

By the time the engine is in this final state, most of the observations are anomalies. The progression of the anomalies to this stage indicates that the engine is deteriorating. This allows the business to decide when the piece of equipment should be repaired or removed from service.

Support Vector Data Description Method

To reiterate, SVDD approaches the problem as follows:

- Create a minimum radius hypersphere around the training data.
- Build the model with kernel functions to add flexibility.
- Compare the value of the distance of an observation from the center with the value of the radius. If it is larger, consider the observation an anomaly.

Support Vector Data Description Results

The SVDD model was trained on the first 25% of the measurements for 30 randomly selected engines in the data set. It was then tested on the remaining 188 engines (Gillespie and Gupta 2017). Four (4) of the 188 engines were randomly selected, and their results are shown in Figure 3.9. The X axis represents the cycle of the flight, and the Y axis represents the distance metric output by the SVDD model.

Figure 3.9: Sample SVDD Scoring Results

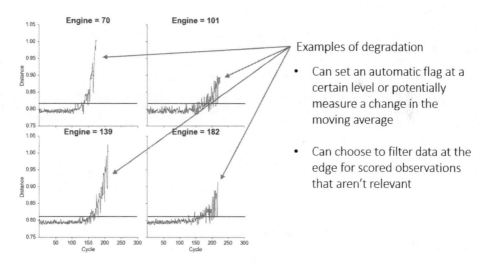

As shown in Figure 3.9, each of the randomly selected engines exhibits a pattern of increasing distance as it gets closer to the end of its life. This pattern begins with some small volatility around a consistent level and then begins to become increasingly volatile as the engine deteriorates. With this knowledge, decisions can be put into place to trigger maintenance or shutdown activities once the distance metric is above a certain threshold or remains above a certain threshold for some time applicable to the use case.

Subspace Tracking

What sets Subspace Tracking (SST) apart from the previously mentioned algorithms is that because it is an online algorithm, it will immediately start training on data upon deployment with no further user intervention required.

SST is a method to detect anomalies and degradation in systems that generate high-frequency, high-dimensional data. SST tracks the principal subspace and principal components over time. It can track changes in the principal subspace, the residual of the principal subspace, and the projection angle between the previous subspace and current observation. SAS has implemented an additional innovation to this technique that monitors the relative or absolute angles of the principal subspace (first principal component).

The following use case demonstrates the detection of parking lot light sensor anomalies at SAS headquarters in Cary, North Carolina. The energy consumption of six floodlights was collected every five minutes. This data contains instances where all lights fail simultaneously due to environmental issues and instances where an individual light fails due to non-environmental issues. One could argue that the latter case is more interesting. Figure 3.10 shows an example of all lights failing simultaneously due to environmental issues. The circled area in Figure 3.11 is an example of an individual light failure.

Figure 3.10: Energy Consumption in Six Parking Lights: All Lights Fail Due to Environmental Issues

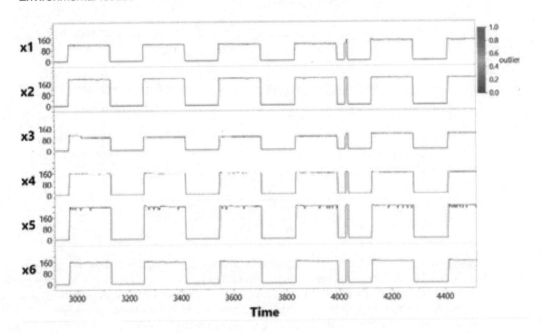

Figure 3.11: Energy Consumption in Six Parking Lights: Fifth Parking Light Shows Different Behavior

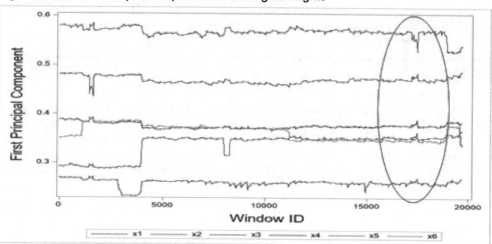

The first principal component of the six lights is shown in Figure 3.12. This plot is useful in identifying which light is failing. For example, at around window 17500, light x5 fails, as can be seen by the sudden decrease of the first principal component in x5. In contrast, the first principal components in the other five variables increase.

Figure 3.12: First Principal Component of Parking Lot Lights

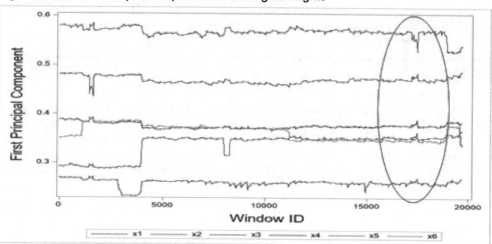

Note, however, if there are many sensors over a long period, then it would be difficult to track all first principal components. Thus, the angle change, shown in Figure 3.13, can help identify the windows of interest. For example, a sudden increase can be seen in the angle change around window 18000. The environmental failure seen in Figure 3.10 is not alerted by a large angle change, as expected.

Figure 3.13: Angle Change of Parking Lot Lights using SST

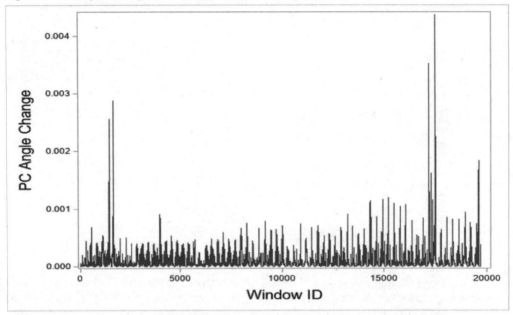

The Facilities department at SAS can use this insight to investigate the failing light and determine the cause of the issue before implementing a plan to restore it. The use cases presented in this chapter demonstrate how to generate actionable insights on streaming data in real-time.

Conclusion

What distinguishes SAS offerings in IoT is the market-leading analytical functionality for streaming data and data at rest and its application in a multi-phase analytics cycle. SAS Event Stream Processing (ESP) can apply advanced analytical techniques to streaming data at the edge, in the fog, and in the cloud. The insights obtained through the application of these analytical techniques can predict failure and anomalies before they become costly or disastrous. They can also detect fraudulent business transactions as they occur, preventing financial disasters.

The advanced analytics supported by ESP is critical to the success of any IoT investment in any domain.

Here are the top findings of a recent survey that articulates the perceived and experienced power of advanced analytics in IoT based on the results of a global study conducted by IDC and sponsored by SAS, with the support of Intel and Deloitte (IDC Study 2019).

1. **Ninety percent of those who are heavily using AI with IoT are exceeding value expectations.** A significant predictor of an organization's ability to deliver value from IoT across an enterprise is the considerable use of AI. The true value of IoT data is realized only when combined with AI and analytics.

2. **Nearly three out of four respondents said the combined value of AI and IoT capabilities exceeded their expectations.** Study respondents across all geographies reported greater success for their IoT initiatives when using AI, with nearly three out of four respondents saying that the value of combined Artificial Intelligence of Things (AIoT) capabilities exceeded their expectations.

3. **Senior leaders already believe the combination of AI and the IoT is strategically important.** Not only are senior leaders overwhelmingly involved with IoT project decisions (79%), but 56% of those senior leaders affirm that AIoT value exceeds expectations.

4. **Companies that use AI and IoT together are more competitive than those using only IoT.** Organizations that have developed an AIoT capacity report much stronger results across several critical organizational goals – from their ability to speed up operations and introduce new digital services to improving employee productivity, decreasing costs, and more.

5. **AIoT capabilities play a considerable role in rapid planning processes, more than previously expected.** Companies that rely on IoT data to inform day-to-day decision-making use it overwhelmingly for operational decisions (68%) rather than planning-oriented decisions (12%). When AI enters the picture, the number of respondents using this data to drive planning-oriented decisions nearly triples, increasing to 31%.

SAS delivers on the promise of advanced analytics impacting business outcomes with the use of streaming analytics.

References

Chaudhuri, A., Kakde, D., Jahja, M., Xiao, W., Jiang, H., Kong, S., and Peredriy, S. (2016). "Sampling Method for Fast Training of Support Vector Data Description." https://arxiv.org/abs/1606.05382.

Gillespie, R. and Gupta, S. (2017) "Real-time Analytics at the Edge: Identifying Abnormal Equipment Behavior and Filtering Data near the Edge for Internet of Things Applications." https://support.sas.com/resources/papers/proceedings17/SAS0645-2017.pdf.

IDC Study. (2019). "AIoT: How IoT Leaders Are Breaking Away." www.sas.com/aiot-study.

SAS. SAS Event Stream Processing 6.2 Documentation. https://support.sas.com/en/software/event-stream-processing-support.html#documentation.

Saxena, A., and Goebel, K. (2008). "Turbofan Engine Degradation Simulation Data Set." Accessed January 17, 2017. NASA Ames Prognostics Data Repository. NASA Ames Research Center, Moffett Field, CA https://ti.arc.nasa.gov/tech/dash/pcoe/prognostic-data-repository/.

Saxena, A., Goebel, K., Simon, D., and Eklund, N. (2008). "Damage Propagation Modeling for Aircraft Engine Run-to-Failure Simulation." In *Proceedings of the International Conference on Prognostics and Health Management, 2008*, 1–9. Piscataway, NJ: IEEE.

About the Contributors

Gül Ege is the Senior Director in R&D responsible for applied analytical components in the SAS IOT division. She leads the analytical component development for high frequency machine data for a variety of IOT domain.. In her 30 years in SAS, she has worked on financial analysis, forecasting, financial risk, retail price optimization, size optimization, and manufacturing vertical solutions. She is a recipient of the SAS CEO Award of Excellence and earned the NCSU-ISE Distinguished Alumni award in 2011. Gül received her PhD in ISE from NCSU.

Recent college graduate **Bennett McAuley** serves as an Associate Technical Writer in Documentation Development. His background in mathematics and data science landed him in the position to support SAS Event Stream Processing after he completed training in the SAS Technical Enablement Academy. Bennett holds a bachelor's degree from North Carolina Central University and plans to pursue a master's degree in statistics.

Chapter 4: Administering SAS Event Stream Processing Environments with SAS Event Stream Manager

By Katja McLaughlin

Introduction

SAS Event Stream Manager is a web-based client that enables you to deploy SAS Event Stream Processing projects into production environments and test environments. You can group assets into deployments and monitor the status of your deployments on an ongoing basis. Furthermore, you can administer your deployments and manage change over time.

In SAS Event Stream Manager, an *asset* is any identifiable part of a SAS Event Stream Processing deployment that SAS Event Stream Manager can monitor or act upon. Examples of assets include SAS Event Stream Processing projects and ESP servers. A *deployment* is a logical grouping of assets as a single unit for life cycle management and monitoring. That is, a deployment is a group of assets that together does something useful.

Monitoring Your SAS Event Stream Processing Environment

Establishing an effective SAS Event Stream Processing environment requires planning You also need to consider how you can administer your environment effectively over its lifetime.

Imagine a scenario where you run a "smart factory" with sensors in important machine components to detect damage early on. In this smart factory, you want to avoid unnecessary costs and protect your investment in the machinery. You set up sensors to gather data about the presence of vibration, noise, temperature, and smoke (Figure 4.1). You know that a component that produces smoke might be only minutes away from breakdown, whereas one that is warmer than expected might be days away from failure. Noise might start weeks before the component breaks down, and you know that vibration is the earliest indicator of forthcoming damage.

Figure 4.1: Indicators of Damage

You gather this data from the sensors that you installed and use SAS Event Stream Processing to analyze the data, using it to perform predictive maintenance and avoid unexpected downtime (Figure 4.2). You use SAS Event Stream Processing Studio to build and test projects to analyze this data. When you start your projects, the results start coming in.

Figure 4.2: SAS Event Stream Processing Analyzes Data Gathered from Sensors

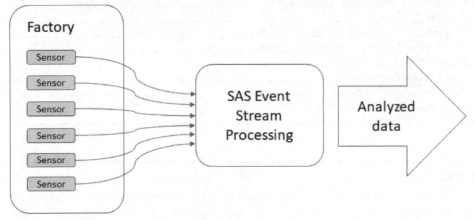

After you have built and deployed a setup that enable you to look after your factory, what happens next? Optimizing the utilization of the machinery, whilst ensuring that you do not overuse it and cause it damage, is something that you must manage over the lifetime of the factory. You do not want to stream data for a few days or weeks – you want to keep monitoring your factory's machine components for months and years to come.

Over time, you might install additional sensors to monitor additional components. Eventually, the existing sensors need to be replaced. You might expand operations and build more factories with yet more machine components and corresponding sensors, which you also want to monitor. Managing a handful of ESP servers is one thing, but in your case, the number of edge devices begins to increase to hundreds and then thousands.

Here is where SAS Event Stream Manager can help. It displays a list of your ESP servers, whether those ESP servers are running on servers located in the cloud, on your premises, or on edge devices (Figure 4.3). When so configured, SAS Event Stream Processing passes information about the existence of ESP servers to SAS Event Stream Manager so that ESP servers automatically appear there.

Figure 4.3: SAS Event Stream Manager Displays ESP Servers from Different Locations

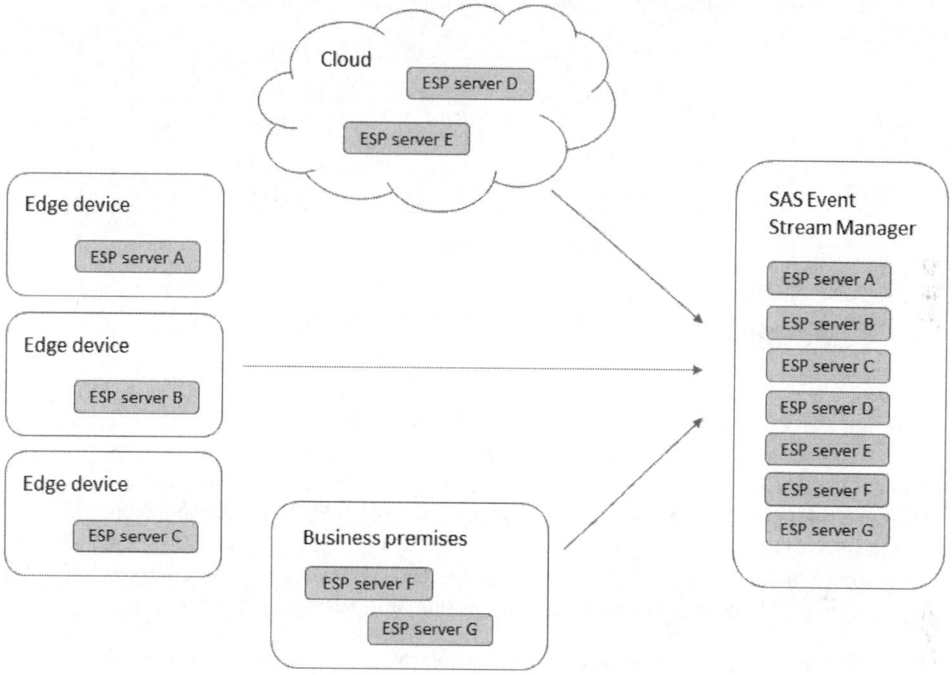

Alternatively, you can also enter ESP server details manually into SAS Event Stream Manager. You can view and edit these ESP server connection details later to adjust how SAS Event Stream Manager connects to those ESP servers. For example, you can change authentication settings or specify that the connection should use SSL.

You can attribute tags to your ESP servers, which can be used to group and filter ESP servers in SAS Event Stream Manager (Figure 4.4). This can prove in useful in your factory, where you have a large number of ESP servers to manage.

Figure 4.4: Examples of Tags

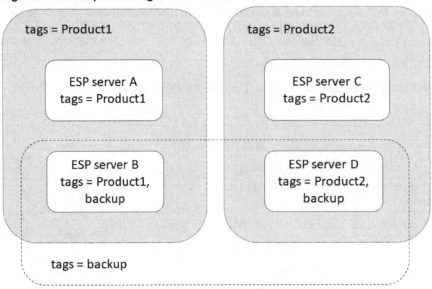

You can also enter a description for each ESP server. For example, you can specify the purpose of the ESP server so that you can differentiate between ESP servers with similar names. In your factory, you might dedicate some ESP servers to testing purposes and record this information in the ESP server description. Without SAS Event Stream Manager, you might have to keep track of your total set of ESP servers and the purpose of each ESP server with spreadsheets or other manual methods. Instead, you can enter those details into SAS Event Stream Manager, where you can tag, group, and filter your ESP servers.

Edge devices can be prone to connectivity problems and power failures, in which case you might not receive the streaming data that you expect. Assume that you have created a deployment called Line 123 in SAS Event Stream Manager to represent a particular production line in your factory. You have added the ESP servers that are running on the edge devices associated with this production line to the deployment. Unfortunately, one of the edge devices has lost connectivity, but you can spot this easily in SAS Event Stream Manager because the deployment's status indicator has changed color. You do not need to be viewing the details for the deployment Line 123 at the time to spot this; the overall view for all deployments displays a status summary for each deployment. Now that you have seen that the deployment Line 123 has a problem, you can drill down to it and find out which ESP servers have problems (Figure 4.5). Then you can take appropriate action.

Figure 4.5: Status Indicators Allow You to Drill Down to Investigate Problems

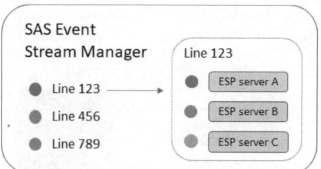

When you drill down to view the details for the deployment Line 123, SAS Event Stream Manager tells you not just whether the associated ESP servers are available but also reveals the status of the projects running on those ESP servers. A full range of status information is available for projects that were executed from SAS Event Stream Manager's web-based interface, as opposed to projects executed directly from SAS Event Stream Processing by using the command-line interface. If ESP servers are available and they do not have any projects in warning or error state, their status is displayed as good. If at least one project running on a particular ESP server is in warning or error state, that ESP server's status reveals that the ESP server is available but there are project errors. By contrast, if an ESP server has at least one project that was not executed from within SAS Event Stream Manager, its status is displayed as "unmanaged." The traffic-light style status indicators give you a quick view of the state of your ESP servers.

Executing Projects from SAS Event Stream Manager

SAS Event Stream Manager enables you to execute projects from its web-based interface, providing a visual way to easily load, start, stop, and unload projects. Quickly loading, starting, stopping, and unloading projects can be particularly useful in test environments (as opposed to production environments) where you might want to try out various projects on different ESP servers. SAS Event Stream Manager enables you to choose multiple ESP servers and load, start, stop, or unload a project on all of them in the same action (Figure 4.6).

Figure 4.6: Select Multiple ESP Servers Then Load and Start the Same Project on All of Them

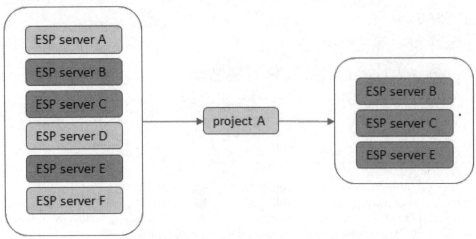

Another benefit of SAS Event Stream Manager is that it enables you to use job templates to automate the execution of projects. A job template is an XML document that is specific to SAS Event Stream Manager that contains a set of instructions to create a job. A job template outlines the steps required to load, start, stop, or unload a project on an ESP server. You can create a job template by using an editor provided within SAS Event Stream Manager.

SAS Event Stream Manager validates your job template content against its job template rules as you write the job template. Job templates make your SAS Event Stream Processing operations easily repeatable. You could compare using job templates to assembling flat-pack furniture. If you follow the instructions closely, you get neatly assembled furniture. If you miss a few steps, your furniture might collapse. Using a job template helps ensure that you do not miss crucial steps in executing your projects.

A job template contains several high-level elements (Figure 4.7).

Figure 4.7: Example of Elements in a Job Template

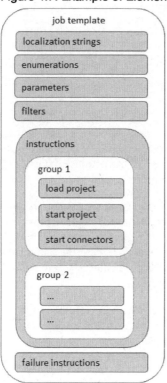

Instructions describe operations that must be performed to create or modify a deployment. For example, you can include instructions to load a project, start a project, start connectors, stop a project, or unload a project. In the smart factory scenario, you could create a job template to load and start a project that receives data from the sensors and processes it. You can also include an instruction to modify an existing running project.

Instructions in job templates can also be grouped together. Grouping instructions can make it easier to manage common dependencies. For example, you could specify that an instruction should not be executed until the previous instruction has been completed. Grouping instructions also enables you to execute a job on several ESP servers.

Here is an example of using instructions in a job template and grouping the instructions. The details of the elements have been removed.

Example Code 4.1

```
<instructions>
    <group <!-- The attributes of the first group element are added
      here.--> >
        <load-project <!-- The details for the load-project instruction
          are added here. --> />
        <start-project <!-- The details for the start-project
          instruction are added here. --> />
        <start-connectors <!-- The details for the start-connectors
          instruction are added here. --> />
```

```
      </group>
      <group <!-- The attributes of the second group element are added
       here. --> >
         <load-project <!-- The details for the load-project instruction
          are added here. --> />
         <start-project <!-- The details for the start-project
          instruction are added here. --> />
         <start-connectors <!-- The details for the start-connectors
          instruction are added here. --> />
      </group>
</instructions>
```

If any of the instructions in your job template fail, you can include *failure instructions* in your job template. Failure instructions can contain the same range of instructions as standard instructions. That is, you can specify to load projects, start projects, and so on.

Job templates can also contain *parameters*, whose purpose is to guide user input. When you deploy a job template, a window appears, and you are prompted to select the deployment against which the job template is to be applied. The parameters element in the job template can specify that a user must also provide further information, such as entering text input or selecting an option from a list. That is, parameters enable a user to enter data into the deployment and customize the job template when it is deployed.

The parameters element includes child elements called *selectors*. In the following job template example, the `project-selector` element enables the user to select from a list of projects stored in SAS Event Stream Manager's repository.

Example Code 4.2

```
<parameters>
   <project-selector id="projectSelector"
    localization-id="projectSelectorLabel"
    required="true"/>
</parameters>
```

The `localization-id` attribute in the above example relates to localization strings, which are discussed later in this chapter.

In the factory example, your job template might require a user to select the project to deploy, the ESP server to deploy the project onto, and the type of machine component to be analyzed. The project is selected from a list of projects stored in SAS Event Stream Manager's repository and the ESP server from a list of ESP servers that are associated with the selected deployment – you do not need to create these lists. The type of machine component is selected from a list of components that is defined by using an *enumeration*. An enumeration is a finite list of options that restricts user input and is referenced from the parameters of the job template (Figure 4.8).

Figure 4.8: Selecting Options for Deploying a Project

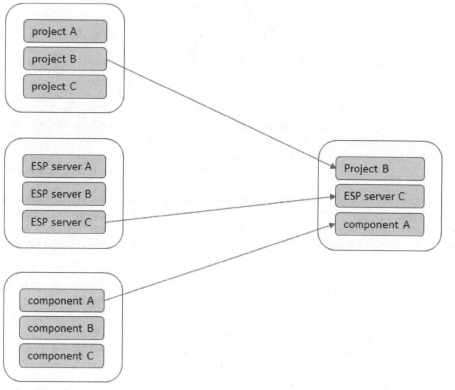

Here is an example of using the `enumerations` element in a job template.

Example Code 4.3

```
<enumerations>
      <enumeration id="components">
           <enumeration-value id="compA" localization-id="compA" />
           <enumeration-value id="compB" localization-id="compB" />
           <enumeration-value id="compC" localization-id="compC" />
      </enumeration>
</enumerations>
```

Localization strings within the job template contain the string values for the labels displayed by SAS Event Stream Manager when a user executes the job template. Each string value is fully localizable within one or more language groups. In the factory scenario, localized values in job templates can be used with a localized version of SAS Event Stream Manager. For example, you might have factory operations in several countries, and SAS Event Stream Manager might be used by staff in these various locations. To ensure standardization, the same job templates can be used across different geographies. The SAS Event Stream Manager user interface reacts to the locale of a user's browser and display the values in the correct language.

Here are some examples of localization strings. Notice how values of the `localization-id` attributes that appeared in previous examples are defined here.

Example Code 4.4

```
<localization-strings default-language="en-us">
   <language id="en-us">
         <string id="projectSelectorLabel">Select a project</string>
         <string id="compA">Component A</string>
      <string id="compB">Component B</string>
         <string id="compC">Component C</string>
   </language>
</localization-strings>
```

The job template for the factory scenario can also contain *filters* for ESP servers. SAS Event Stream Manager enables you to search for ESP servers that match certain criteria and save those criteria as a filter that you can reference in job templates. The filter resolves to the ESP server selected by the SAS Event Stream Manager user when the job template is deployed.

For example, you might want to deploy a project to ESP servers running on edge devices of a specific type, to which you have assigned the tag "SensorTypeA." If the number of edge devices with the tag "SensorTypeA" changes in the future, a filter that references this tag still finds all matching ESP servers. The filter expression language is straightforward; you simply need to specify `tags='SensorTypeA'`.

Furthermore, you can use job template parameters to enable a SAS Event Stream Manager user to select a filter, rather than hardcoding the selected filter into the job template. In this case, a separate window is displayed to users, enabling them to choose the desired filter. For example, in your factory you could enable a user to select whether to deploy a job template to ESP servers with the tag "SensorTypeA," or to ESP servers with the tag "SensorTypeB," and so on.

In general, creating job templates is most beneficial in complex situations; for example, when you want to collect user input and customize the job template when it is deployed, or if you want to use filters. The purpose of job templates is to automate the execution of projects. On the other hand, executing projects by using the user interface controls is quick and easy, and sufficient in many straightforward situations.

Whether you deploy a job template or select a user interface control to load, start, stop, or unload a project, the result is that a job is created. A job is a set of tasks to be executed by SAS Event Stream Manager on various assets associated with one or more deployments (Figure 4.9).

Figure 4.9: Deploying a Job Template

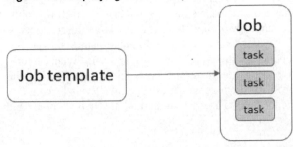

SAS Event Stream Manager displays the status of active and historical jobs. You can use this information to help resolve problems with your deployments. The information displayed for each job can include:

- job status
- current job progress as a percentage of the total, start, and end times
- job duration
- the name of the deployment to which the job relates

You can sort this information in various ways. For example, you could view information for all jobs started today. You can drill down into job details where SAS Event Stream Manager provides information about the tasks that make up the job.

In the smart factory example, you might have a job template that includes instructions to load and start a project and then start publish/subscribe connectors to collect data from the sensors (Figure 4.10). SAS Event Stream Manager tell you which tasks were completed and whether tasks were completed on all relevant ESP servers. For example, SAS Event Stream Manager might reveal that the publish/subscribe connector failed to be started on one of the ESP servers where the job template was deployed. We do not merely know that the job was not completed successfully. Instead, we have a starting point for our investigation into the matter.

Figure 4.10: Monitoring a Job

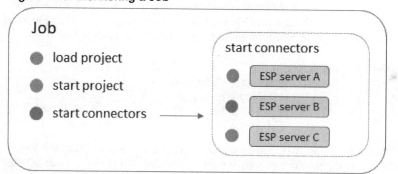

Governing and Testing Assets

In addition to keeping track of jobs, SAS Event Stream Manager helps you handle the journey of your deployments, projects, and job templates from development and test through to the production environment.

When you create a project in SAS Event Stream Processing Studio and publish it, the project automatically becomes available in SAS Event Stream Manager. This helps ensure that when you execute projects from SAS Event Stream Manager, you do not accidentally deploy an outdated version of a project. When you deploy job templates or load or start projects using the user interface controls, SAS Event Stream Manager defaults to the latest version of the job template, but also enables you to select older versions if required.

SAS Event Stream Manager gives you access to the details for previous versions of your projects. You can open a previous version and view all the same information as for the latest version of the project. This information includes details such as the project diagram, the XML that makes up the project, and project metadata.

SAS Event Stream Manager enables you to mark deployments, project, and job templates as production assets to prevent assets meant for testing from accidentally being used for production that is, being used in a live environment. You can mark these assets as production assets when you create them. You can change the production status to a different status later. For example, you can change a production project to a non-production project, or you can change a non-production project to a production project. You can deploy a non-production job template only against a non-production deployment. However, in this case, SAS Event Stream Manager permits you to select either a production project or a non-production project to deploy. This choice enables you to not only test your current production projects and job templates, but also test non-production projects and job templates to assess whether they are suitable for being marked as production assets in the future. The following example illustrates this.

In your smart factory scenario, you already have an existing setup to analyze sensor data. As previously noted, you have a deployment called Line123. Assume that thus far you have used a project named SensorProjectA and a job template named VibrationTemplateA. You use VibrationTemplateA to analyze data obtained from the sensors to determine whether vibration is at a level that require a maintenance engineer to service or repair a machine component. You currently use these three assets for production purposes and have marked them as production assets in SAS Event Stream Manager (Figure 4.11).

Figure 4.11: Asset Example – Step 1

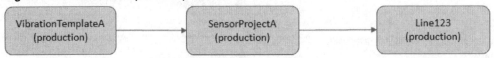

Assume now that you have been working on an improved project and an improved job template to make enhancements to how sensor data is analyzed. You have created another deployment called Line123-Test. This deployment is marked as a non-production deployment. The ESP servers that you have associated with this second deployment are ESP servers that you want to use for testing. Because you have this second deployment, you can keep the improved project and job template separate from the production environment.

The improved project is called SensorProjectB. You have created it in SAS Event Stream Studio and published it so that it is now available in SAS Event Stream Manager. You have also created an improved job template and called it VibrationTemplateB. Furthermore, you have marked SensorProjectB and VibrationTemplateB as non-production assets. You then test SensorProjectB and VibrationTemplateB against the deployment Line123-Test (Figure 4.12).

Figure 4.12: Asset Example – Step 2

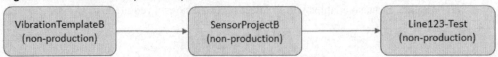

Suppose you conclude that SensorProjectB is ready to be promoted to production but VibrationTemplateB is not. You want to start using SensorProjectB as soon as possible, so whilst you continue working on improving VibrationTemplateB further, you want to check if you could use SensorProjectB straight away. To do this, you can deploy VibrationTemplateA with SensorProjectB against Line123-Test to check that SensorProjectB works well with the original job template VibrationTemplateA. That is, you can deploy the original job template with the improved project to our test environment (Figure 4.13).

Figure 4.13: Asset Example – Step 3

If you tried to deploy VibrationTemplateB against Line123 (the production environment), SAS Event Stream Manager would not permit it. This is because VibrationTemplateB is marked as a non-production job template, and Line123 is marked as a production deployment.

Suppose you then continue work on VibrationTemplateB and make some further improvements to it. To test it, you deploy VibrationTemplateB using SensorProjectB (which is now a production asset) against Line123-Test – that is, the test environment (Figure 4.14).

Figure 4.14: Asset Example – Step 4

When you are happy with VibrationTemplateB, you can mark it as a production asset and deploy it using SensorProjectB against Line123 – that is, against the production environment (Figure 4.15).

Figure 4.15: Asset Example – Step 5

These examples show how you can test a combination of production and non-production job templates and projects against a non-production environment. You initially deployed VibrationTemplateB with SensorProjectB against Line123-Test whilst all three assets were marked a non-production asset. However, later you marked SensorProjectB as a production project and deployed VibrationTemplateB using SensorProjectB. You were able to do this even though SensorProjectB had already been marked as a production project because you were deploying VibrationTemplateB to a non-production deployment, Line123-Test.

84 *Intelligence at the Edge*

Handling Changes to ESP Servers

As your SAS Event Stream Processing environment evolves, you might need to move ESP servers in and out of deployments. SAS Event Stream Manager provides a clear view of all ESP servers that are currently not assigned to any deployment (Figure 4.16). When you view a specific deployment, you can remove ESP servers from it and assign other ESP servers to it. Before you can remove an ESP server from a deployment, you must stop and unload any projects that were previously running on the ESP server.

Figure 4.16: SAS Event Stream Manager Shows ESP Servers Assigned to Each Deployment, and Unassigned ESP Servers

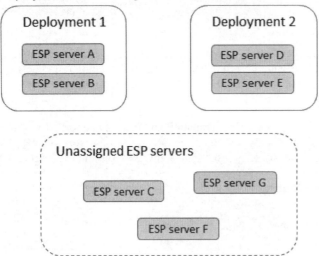

Recall that SAS Event Stream Manager displays a list of your ESP servers, whether those ESP servers are running on servers located in the cloud, on your premises, or on edge devices. SAS Event Stream Processing can pass information about the existence of ESP servers to SAS Event Stream Manager so that ESP servers automatically appear in SAS Event Stream Manager. You can also enter ESP server details manually into SAS Event Stream Manager.

SAS Event Stream Manager provides information about memory use for each ESP server. Furthermore, it reveals CPU use for each project and window. This information can help you investigate problems with ESP servers further.

SAS Event Stream Manager also displays the version of SAS Event Stream Processing that each ESP server is running with.

You can also use SAS Event Stream Manager to monitor your metering servers. This provides you metrics about the number of events that your system processes. It also ensures that your production ESP servers are in compliance with the terms of your software license. SAS Event Stream Manager provides a breakdown for each license, host, and ESP server. You can export detailed information about events to a comma-separated values (CSV) file.

Integrating with SAS Model Manager

SAS Model Manager stores SAS and open-source analytical models. You can use SAS Model Manager to develop candidate models and compare candidate models for champion model selection. You can then use it to publish and monitor champion and challenger models to ensure optimal model performance.

SAS Event Stream Manager works together with SAS Model Manager. Assume that in your smart factory scenario, the project SensorProjectB references models that are stored in the SAS Model Manager common model repository. When you use SAS Event Stream Manager to execute SensorProjectB, the model is retrieved from the SAS Model Manager common model repository and written to the ESP server. SAS Micro Analytic Service modules are used to accommodate such models (Figure 4.17).

Figure 4.17: A Model is Retrieved from SAS Model Manager

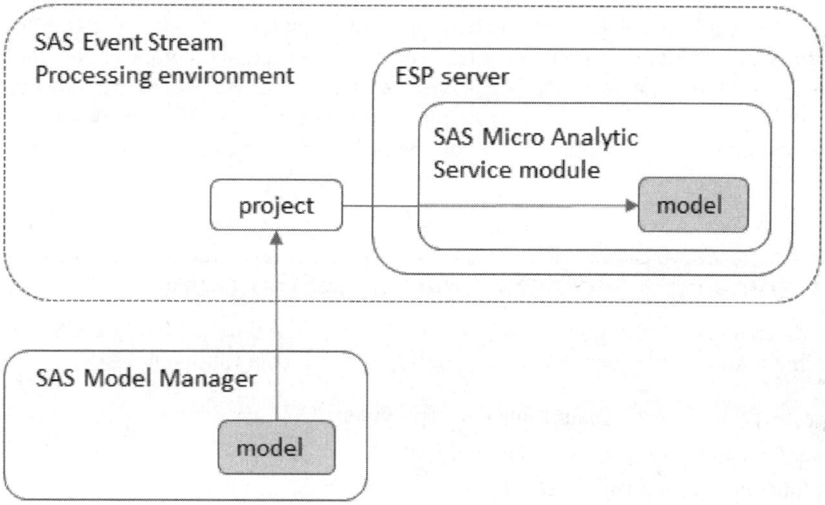

Later, a new champion model is declared in SAS Model Manager. SAS Event Stream Manager automatically displays a notification to let you know that you can update the running project SensorProjectB to reflect updates to the model. If you have more than one project that references the model, you can update some or all of them to use the new version of the model. SAS Event Stream Manager runs a job to fetch and deploy the new champion model for the selected projects. Edge data from our machine component sensors is now being analyzed using this new champion model.

SAS Event Stream Manager updates the version information for your project to reflect the new model. For example, SensorProjectB might be updated from version 1.2 to version 1.3. The new version also automatically becomes available in SAS Event Stream Processing Studio.

Accommodating Different User Roles

Personnel with various roles in the organization can work together in SAS Event Stream Manager, even if there is overlap in their roles and the tasks that they need to perform.

For example, when you are setting up your SAS Event Stream Processing environment, development staff (such as application developers) might use SAS Event Stream Manager to check that the relevant ESP servers appear in SAS Event Stream Manager and to register any further ESP servers.

Development staff or operations staff (such as IT engineers, data scientists, or system architects) might use SAS Event Stream Processing Studio to create and publish projects, and then use SAS Event Stream Manager to create job templates. They might also deploy these job templates in a non-production environment to test them. Following this, operations staff might promote test assets to production assets.

Assigning ESP servers to deployments and deploying job templates can also be a task for operations staff. Alternatively, it could be a task for business users who monitor and control the status of the ESP environment after it has been set up. SAS Event Stream Manager gives you the flexibility to allow business users to govern and control the ESP environment solely through the web interface – without using a command-line interface or needing scripting skills.

Example: Deploying a Project Using a Job Template

The following example, which you can follow step-by-step if you have SAS Event Stream Processing and SAS Event Stream Manager, shows how to do the following tasks:

- create a SAS Event Stream Manager deployment
- associate an ESP server with the deployment
- upload a project and a job template
- deploy the job template
- monitor the deployment
- stop a running job

The example uses the files listed below. These files are provided for you in the SAS Event Stream Manager examples package that you can download from https://support.sas.com/kb/60/324.html.

- trades_connector_stocksymbol_job_template.xml is a job template that loads and starts a project. The job template also collects user input. When the job template is deployed, a placeholder is replaced with the stock code that is selected by the user.
- trades1M.csv contains stock trade data used as input events in this example.
- placeholder_filtered_trades.csv is an output file. When you deploy the job template, the project processes stock trades and filters out trades that match a specified stock code. The project then writes these trades into the placeholder_filtered_trades.csv file.

- stop_project_job_template.xml is a job template that you can deploy to stop the running project when you have finished exploring this example

The project used in the example is trades_connector_stocksymbol_project.xml. The project contains three windows (Figure 4.18):

- The Source window (`source_win`) is where trades data from the trades1M.csv file enters the model.
- The Filter window (`symcode_filter`) contains a filter expression that identifies events with a specified stock symbol code.
- The Aggregate window (`aggFromFilt`) places events into aggregate groups and calculates the weighted average price, maximum price, minimum price, and standard deviation. This window also writes the results in the placeholder_filtered_trades.csv file.

Figure 4.18: Continuous Query for Evaluating Stock Trades

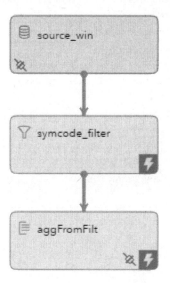

The project processes stock trades and filters out trades that match a specified stock code. The project then writes these trades into an output file named placeholder_filtered_trades.csv.

Note: To use this example, you need an ESP server that is not in a cluster. You cannot deploy job templates to an ESP server that is in a cluster. The project file that is intended for use with the job templates in the example.

Prepare the Example Files for Use

1. Download the SAS Event Stream Manager examples package from https://support.sas.com/kb/60/324.html.

2. Save the following files that are contained in the package to a temporary location on your computer:

    o   trades_connector_stocksymbol_project.xml

    o   trades_connector_stocksymbol_job_template.xml

    o   stop_project_job_template.xml

3. Download the SAS Event Stream Processing code samples package from https://support.sas.com/downloads/package.htm?pid=2421.

4. Save the /xml/vwap_xml/trades1M.csv file from the package to the server on which SAS Event Stream Processing is running.

5. Update the trades_connector_stocksymbol_project.xml file to point to the location of the trades1M.csv file:

    a. Open the trades_connector_stocksymbol_project.xml file in a text editor.

    b. Locate the following line (for example, by searching):
        ```
        <![CDATA[/data/input/trades1M.csv]]>
        ```

    c. Update the directory path to point to the location of the trades1M.csv file on the server on which SAS Event Stream Processing is running.

    d. Locate the following line (for example, by searching):
        ```
        <![CDATA[/data/placeholder_filtered_trades.csv]]>
        ```

    e. Update the directory path to point to a location on the server on which SAS Event Stream Processing is running where the output file can be written.

    f. Save the trades_connector_stocksymbol_project.xml file.

Create the Stock Trade Deployment

1. Open SAS Event Stream Manager

2. On the **Deployments** page, click .
    The Deployment Properties window appears.

3. In the **Name** field, enter "Stock Trade".

4. In the **Description** field, enter "A deployment for processing stock trades".

5. In the **Tags** field, enter "stock".

6. Click **OK**.

The Stock Trade deployment appears on a new page.

Add an ESP Server

Now you associate an ESP server with the Stock Trade deployment.

Note: Ensure that you have an ESP server that you can use for this example. To follow this example, you need an ESP server that is not in a cluster. You cannot deploy job templates to an ESP server that is in a cluster. You can view available ESP servers that are not in a cluster on the **Unassigned Servers** page. SAS Event Stream Manager detects some ESP servers automatically. You can also connect directly to an ESP server. In this case, SAS Event Stream Manager becomes aware of the ESP server only after you have connected.

To add an ESP server to the Stock Trade deployment:

1. Click and select **Add an unassigned server**.
    The Add and Remove ESP Servers window appears.
2. Move the desired ESP server from the **Available servers** table to the **Selected servers** table.
3. Click **OK**.

The ESP server appears in the table on the Stock Trade deployment page.

Upload and Publish the Stock Trade Project

Note: You make the stock trade project available to SAS Event Stream Manager by using SAS Event Stream Processing Studio to upload and publish the project. However, if SAS Event Stream Processing Studio is running as a stand-alone application, you cannot use it to publish projects. Therefore, you must upload the stock trade project to SAS Event Stream Manager.

To use SAS Event Stream Processing Studio to upload and publish the stock trade project:

1. Upload the trades_connector_stocksymbol_project.xml file to SAS Event Stream Processing Studio.
    The filteredtrades project appears on the **Projects** page in SAS Event Stream Processing Studio.
2. Publish the filteredtrades project.

Afterward, the filteredtrades project appears on the **Projects** page in SAS Event Stream Manager.

To explore the project's contents, right-click the filteredtrades project on the **Projects** page and select **Open project**. The project opens, and you can click the **Diagram**, **XML**, **Details**, **Versions**, and **Files** tabs to explore their contents.

Upload the Stock Trade Job Template

1. On the **Job Templates** page, click ⋮ and select **Upload job template**.
   The Upload Job Template window appears.
2. Fill out the fields as follows:
   ○ **File**: Navigate to the trades_connector_stocksymbol_job_template.xml file.

   ○ **Tags**: Enter "stock".

   ○ **Production template**: Leave this check box deselected.

   ○ **Version notes**: Enter "First version".

3. Click **OK**.

The Filter Trades by Stock Code job template appears on the **Job Templates** page.

Note: To explore the job template's contents, right-click the Filter Trades by Stock Code job template and select **Open**.

Deploy the Stock Trade Job Template

1. On the **Job Templates** page, select the Filter Trades by Stock Code job template and click ⬒ .
   The Job Template "Filter Trades by Stock Code" v.1 window appears.
2. Fill out the fields as follows:

   ○ **Deployment**: Select **Stock Trade**.

   ○ **Project**: Select **filteredtrades**.

   ○ **Latest available version**: This field is automatically set to 1.0.

   ○ **ESP server**: Select your ESP server.

   ○ **Stock symbol code**: Select **Micro Focus**.

   When the job template is deployed, trades for Micro Focus stock are identified with the MCRO stock code.
3. Note: The **Deployment** field always appears in this window. The other fields appear because they are specified in the job template parameters.
4. Click **OK**.

The **Filter Trades by Stock Code** page appears. When the Load Project task has completed successfully, its status is displayed as ✓ .

Monitor the Deployment

1. On the **Deployments** page, select the Stock Trade deployment and click .
   The **Stock Trade** page appears.
2. In the main table, select your ESP server.
   The MCRO_1 project appears in the bottom pane.
3. Right-click the MCRO_1 project and select **Open running project**.
   The **MCRO_1** page appears.
4. Select each tab in turn to view the results for each window:

   ○ The **source_win** tab displays a snapshot of input events in real time.

   ○ The source_win window has no index. After all the events in the CSV file have been displayed, the table is empty.

   ○ The **symcode_filter** tab lists trades for the selected stock code, MCRO.

   ○ The symcode_filter window has an index. After all the events in the CSV file have been displayed, the table still contains some events.

   ○ The **aggFromFilt** tab displays aggregated results.

   ○ The aggFromFilt window has an index. After all the events in the CSV file have been displayed, the table still contains some events. However, to see all the results, open the placeholder_filtered_trades.csv file on the server on which SAS Event Stream Processing is running (Figure 4.19).

Figure 4.19: Contents of the Placeholder Filtered Trades CSV File

	A	B	C	D	E	F	G
1	I	N	MCRO	26.97	26.97	26.97	0
2	UB	N	MCRO	26.97	26.97	26.97	3.89E-07
3	D	N	MCRO	26.97	26.97	26.97	0
4	UB	N	MCRO	26.97	26.97	26.97	0
5	D	N	MCRO	26.97	26.97	26.97	3.89E-07
6	UB	N	MCRO	26.96099	26.84	26.97	0.058138
7	D	N	MCRO	26.97	26.97	26.97	0
8	UB	N	MCRO	26.95721	26.84	26.97	0.052313
9	D	N	MCRO	26.96099	26.84	26.97	0.058138

If you want to see results for the other two stocks, deploy the project again and select a different stock.

Stop the Stock Trade Job

After you have finished exploring this example, you can stop the running job by deploying another job template file, stop_project_job_template.xml. Deploying this job template stops the selected project and unloads it from the ESP server.

1. On the **Job Templates** page, click ⋮ and select **Upload job template**.
    The Upload Job Template window appears.
2. In the **File** field, select the stop_project_job_template.xml file.
3. Click **OK**.

4. Select the Remove a Running Project v.1 job template and click 🡒 .
    The Job Template "Remove a Running Project" v.1 window is displayed.
5. Fill out the fields as follows:

    ○   **Deployment**: Select **Stock Trade**.

    ○   **ESP server**: Select your ESP server.

    ○   **Project to unload**: Select **MCRO_1**.

    Note: If you selected a different stock code when exploring the example, select the code for that stock instead.
6. Click **OK**.

The **Remove a Running Project** page appears.

This job template has two tasks: Stop Project and Unload Project. When each task has completed successfully, the status for each job is displayed as ✓ .

Conclusion

The proliferation of data from the edge makes it essential to manage ESP servers flexibly. But not all ESP servers are the same. They run different projects, different versions of software, deploy different analytical models, and so on. SAS Event Stream Manager provides a repeatable, automated, and traceable process to monitor and govern groups of ESP servers. It works on premises, on the edge, and in the cloud. It can display alerts when new champion analytical models are available, and it can update ESP servers with those new champions. With an easy-to-use graphical interface, you can quickly get SAS Event Stream Processing projects up and running.

About the Contributor

Katja McLaughlin is a Senior Technical Writer at SAS working on SAS Event Stream Manager. She collaborates with stakeholders such as product managers, UI designers, developers, testers, and customer support staff to create online Help documentation. She also provides guidance to the development team to improve user interface text. Katja holds a degree from the University of Glasgow in Scotland and is a member of the Institute of Scientific and Technical Communicators.

Chapter 5: SAS Event Stream Processing in an IoT Reference Architecture

By Steve Sparano, Shawn Pecze, Prasanth Kanakadandi, and Byron Biggs

What is an IoT Reference Architecture?

A *reference architecture* provides a template architecture for a specific domain and a common vocabulary to use during implementations. An Internet of Things (IoT) reference architecture provides a template to implement a complete, robust, and extensible IoT solution. It includes technologies for the following:

- analytical model development
- model management
- streaming project development
- performance reporting
- deployment

Such a reference architecture provides capabilities to develop, test, deploy, monitor, and update all components in a real-time system. Each component must interoperate with the others to deliver business insights and decisions in real-time. The entire system must ensure that the analytical models in use are the most current.

But an IoT reference architecture cannot simply be a collection of disjointed tools. It should exhibit the characteristics detailed in Table 5.1.

Table 5.1: Characteristics of an IoT Reference Architecture

Characteristic	Description
Open	Interoperates with an ecosystem of technologies, including message buses, databases, containers, and open-source languages for streaming analytics.
Multi-phase analytics	Integrates scoring models that are trained offline and deploys them in-stream. Trains and scores events in stream. Delivers models and events from the edge to the enterprise with deployment capabilities.
Integrated	Integrates analytic development, model management, model deployment, performance monitoring, model updates, and deployment across the analytical life cycle.
Self-improving feedback mechanism	The distribution of incoming data values drifts over time as assets (for example, machines, people, buildings, vehicles) age. Scoring algorithms are monitored and, when needed, updated with the most recent score code.
Automation	Keeping the numerous analytical models up-to-date requires automation for retraining and updating the models. Model deployment, whether on the edge or in the cloud or both, is automated.
Collaborative	Data scientists and operations teams collaborate on the following: • What systems are most critical? • What needs to be predicted and when? • Which actions need to be prescribed? • How can you quickly access results to make rapid decisions?

IoT Challenges

Table 5.2 lists the challenges that must be met in the realm of streaming analytics and real-time event processing in order to successfully deliver an effective IoT solution.

Table 5.2: IoT Challenges

Challenges	Description
Data types	Machine data is typically structured. Significant business value is locked away in unstructured data such as the following: • call center transcripts • system notes • audio files • video • image data • social media streams where customers share their feedback on products and services Various types of data need to be ingested to ensure all of this value is extracted for maximum benefit
Speed of data movement	Systems must be able to ingest data at increasing speeds. They should not simply store it but analyze it, score it, aggregate it as the data arrives. They cannot perform these actions at the cost of throughput or latency.
Variety of development languages	Data scientists work with a variety of programming languages and open analytic frameworks across different phases of the analytic life cycle. These frameworks need to be integrated into the streaming analytics platform.
Cross-platform independence	From the edge to the cloud, platforms change depending on the task, memory footprint, processing power, chip architecture, and disk footprint. The variety of platforms whether they are edge or cloud, require support for monitoring, deployment, and execution of analytical models. A reference architecture needs to accommodate this variety of platforms by maintaining an independence from their specific characteristics and provide open APIs that these platforms can use for integration.

IoT Reference Architecture Components

Like any complete and robust reference architecture, there are a core set of technology components needed for real-time streaming data analysis. These components deliver the capabilities listed in Table 5.1 and address the challenges of speed and data variability. They offer a way to operationalize streaming analytics. The requisite tools and capabilities needed are outlined in Figure 5.1.

Figure 5.1: Components of an IoT Reference Architecture

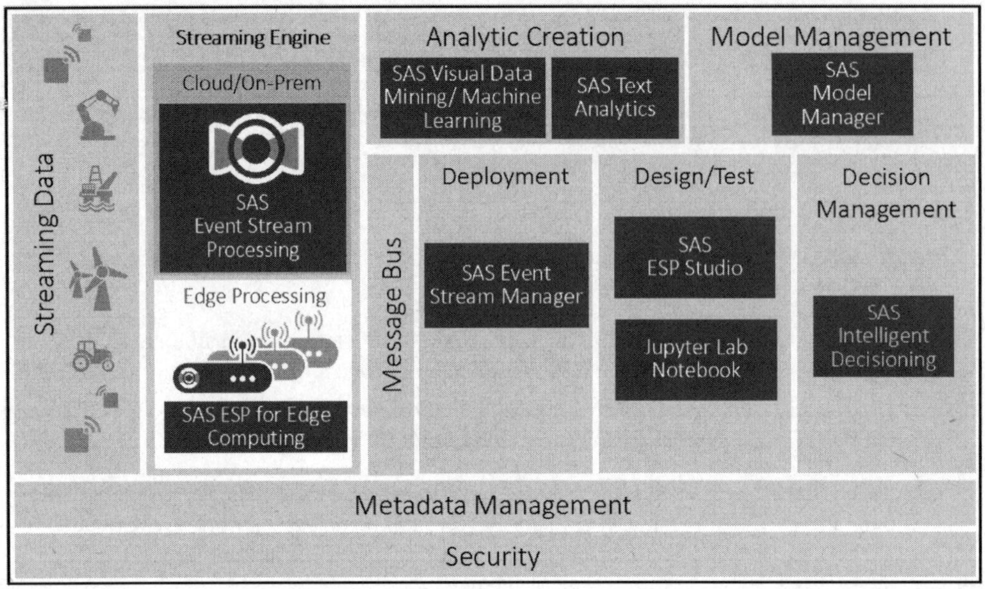

Each component listed provides a set of capabilities required for complete streaming analytics IoT reference architecture (Table 5.3).

Table 5.3: IoT Components

Component	Description
Streaming Engine: Edge Processing	SAS Event Stream Processing for Edge Computing extends the power of the SAS ESP Server to run on edge platforms, which are deployed near the places where IoT data originates: on gateways, in the fog, at aggregation points, in cameras, and in edge assets. The technology delivers the power of a fully functional streaming analytics engine but on a smaller footprint for edge applications. Examples include routers or gateways in predictive maintenance asset monitoring use cases where unlike in the cloud or in data centers, compute, and memory is not virtually unlimited. The Edge Computing environment also provides a more secure setting on a lower cost platform. Costs are lower because data can be processed on the edge, and only events of interest are transmitted over expensive transmission networks and stored in the cloud.
Streaming Engine: Cloud/On Premises	The ESP Server is at the heart of the operational systems executing streaming analytics to analyze streaming events in real time and operate in the cloud or on-premises. The analytical models and ESP projects are published to the ESP Server where they execute in memory and continuously analyze data for patterns or events of interest. A server continuously processes always-on streams of data 24/7. These systems process events with high throughput and extremely low latency.

Component	Description
Analytic Creation	SAS provides tools for analytic model creation including those listed in Figure 5.1. Prominent among them is SAS Visual Data Mining and Machine Learning, which supports the end-to-end data mining and machine learning process with a comprehensive visual and programming interface. All of these tools support data exploration, analytic model creation, model tuning, testing, and publishing. They integrate with a variety of open-source languages to create best-of-breed streaming analytic scoring algorithms for integration with SAS Event Stream Processing projects. For example, businesses might have investments in open-source languages as well as SAS. By providing a means to leverage their existing analytic investments, staff who have already been trained on specific technologies can continue with their familiar environment. This staff is not required to eliminate a proven approach. Various technologies can be used together to deliver complementary technologies for more robust results.
Model Management	SAS Model Manager delivers model management capabilities built on a common model repository with a single location for model storage. This repository supports the collection of performance results from SAS Event Stream Processing (SAS ESP) scoring operations on the edge or in the cloud. This performance reporting can alert analysts to model decay or variable drift. SAS Model Manager provides version control and model publishing and tracking to deliver model governance. SAS Model Manager integrates with SAS Event Stream Manager for event-driven analytical model deployment through message bus technology. This technology automatically delivers alerts to operations teams when more current and refreshed models are available. For example, when model performance drifts due to changes in input data, data scientists can be alerted to these changes. These changes can trigger retraining of models and ensure that the most accurate and current models are applied. This environment needs to track where the models are deployed to enable lineage for delivering updates to the servers executing the models on the edge or cloud or on-premises.
Deployment	Deployment to operational systems includes the ability to publish to a host of IoT run-time servers in the cloud, on-premises, as well as on the edge. SAS Event Stream Manager supports the ability to deploy, monitor, and track health and lineage for ESP projects; automatically update for new Champion models; and collaborate with data scientist for event-driven analytical model deployment. Once deployed, SAS ESP can leverage message bus architecture such as Kafka to deliver resilient, fault-tolerant systems for IoT applications.

Component	Description
Design and Test	SAS Event Stream Processing Studio provides a graphical tool to enable designers to quickly design, test, and validate streaming projects. This speeds development time for the business analyst who can use easy drag-and-drop tools that integrate process flow logic and analytics. SAS Event Stream Processing Studio provides the designer with tools that integrate the power of streaming analytics and facilitates testing of those streaming analytics projects. It can connect to test data sources from a variety of data types. After testing is complete, version management for the streaming analytics projects can be used to track progress. And the design time environment needs to integrate with tools such as SAS Event Stream Manager for rapid deployment to running production servers.
Decision Management	SAS Intelligent Decisioning (SAS ID) provides the ability to act when conditions are detected with SAS ESP. SAS ID provides business rules management integrated with advanced analytics and the ability to design and test business rules and decisions before deployment to production operational systems. Predicting the probability of a fault, a customer's next best action, or detecting and classifying objects is an important part of any analytically driven IoT system. Also, how do you decide what action to take when these events of interest are equally important? For example, after a customer's behavior is identified, how can you ensure they are presented with the right offer out of all possible offers? What is the best action to take when a potential fault is identified for a critical asset in your factory? Decision management ensures you integrate business rules and decision logic with the streaming analytics.
Metadata Management	This includes the ability to capture and manage the data about servers, topology, analytical models, deployments, and streaming projects, among other items in the IoT reference architecture. Metadata facilitates the discovery of objects such as streaming project output schemas or required input schemas for analytical models. Information is made available through open APIs for searching and use when developing analytics. For example, by having information available about streaming analytics projects, other applications can understand what data that project consumes and produces. This makes it easier to understand how to integrate projects.

Component	Description
Security	It is critical to protect devices and data on the edge. Device hacking, unauthorized access, malware, and other unwanted intrusions can jeopardize the operation of edge devices. As data centers have become more hardened over time, edge devices have struggled to keep up. As the number of devices deployed continues to escalate, the risk for unwanted access and possible tampering increases. With edge processing, the hacking on transmission of data can be minimized with data being processed on the edge. Only the necessary events of interest can be relayed for further processing. This does not eliminate the need for secure techniques such as encryption, but it lowers the amount of data required to be transmitted and the risk for data theft as well.

Deployment Considerations

Deployment is where the rubber meets the road. With an IoT Architecture, it's where the analytics are put to work to drive decisions to improve business. A streamlined process that is efficient and capable for supporting device platforms including analytic run-time environments is critical to putting the analytics to work.

SAS Event Stream Manager (ESM) supports a flexible approach that delivers streaming analytics to the edge as well as the cloud using automation. Automation helps deliver streaming projects to large numbers of edge devices and does so in a way that expands as the number of devices in production also expands. Through the management of a collection of logically related servers, called deployments, ESM supports job templates (user-written scripts) to automate actions on the collections of servers.

The operations on the collections of servers ranges from loading projects, unloading projects, and managing data stream using connectors. ESM can perform these operations using parameterized values that keeps the process flexible as streaming analytics projects are deployed.

The capabilities of SAS Event Stream Manager are covered in detail in Chapter 4.

Edge Technologies

When deploying to edge platforms, various edge architectures and technologies require different approaches depending on the use case. SAS ESM supports deployment to different device manufacturers to publish projects to ESP servers running on the edge. For example, SAS Event Stream Manager supports deployments to our customers' proprietary edge devices through use of the built-in ESP for Edge Computing APIs. These APIs enable event stream processing projects to be deployed directly to the device.

The deployment process from ESM uses secure mechanisms to deploy to the edge where processing can be more secure. Event processing on the edge can eliminate the need to transmit all events back to the cloud by detecting events of interest on the edge. Detection takes place in a self-contained streaming analytics environment. This approach supports transmitting only the events of interest, or anomalies, back to the cloud for processing, thereby reducing the data that could be compromised during transmission.

Cloud Technologies

SAS Event Stream Processing is cloud-ready for large-scaled distributed services. With SAS Event Stream Processing 6.2, pre-built containers are available for deployment in a Kubernetes cluster. One container provides a Kubernetes operator for ESP. This operator provides a way to build, drive, and automate elastic ESP servers on the cluster. It enables the preservation of ESP server state and more nimble server management. Other containers provide the following:

- SAS Event Stream Processing Studio
- SAS Event Stream Processing Streamviewer
- SAS Event Steam Manager
- The SAS Event Stream Processing metering server, which aggregates event counts based on license, source window, and hour of the day.

Work is underway to integrate SAS Event Stream Processing with Microsoft Azure at the edge. Sensors would send data in whatever format they use to SAS Event Stream Processing on the Edge. SAS Event Stream Processing would communicate with Azure IoT Edge Runtime, which would connect directly to the Azure Cloud.

Use Case

The following use case presents a typical scenario that demonstrates an IoT reference architecture in action. It trains a model using offline data in SAS Visual Data Mining and Machine Learning, deploys the model into SAS Event Stream Processing, and manages the model with SAS Model Manager. The data set for the example is the Turbofan Engine Degradation Simulation Data Set from the NASA Ames Prognostics Data Repository (Saxena and Goebel 2008). The target variable for the example is FuelRatio, which is an interval target.

Using SAS Visual Data Mining and Machine Learning to Build a Model

SAS Visual Data Mining and Machine Learning enables you to build complex analytics pipelines to determine the best model for your data. You can create SAS Visual Data Mining and Machine Learning pipelines in two ways.

One way to create a pipeline is to use a model built in SAS Visual Analytics as a starting point. Click **Create pipeline** in the report that you have created as shown in Figure 5.2. You can add this pipeline to a new project (SAS Visual Data Mining and Machine Learning project), or you can add the pipeline to an existing project (SAS Visual Data Mining and Machine Learning project).

Figure 5.2: Creating a Pipeline from SAS Visual Analytics

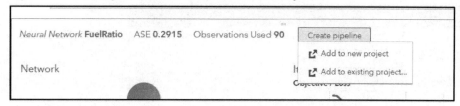

You can also manually create a project in Model Studio and build one or more pipelines. The remainder of this section describes this method of creating a pipeline.

To manually create your project in Model Studio, begin by creating a new SAS Visual Data Mining and Machine Learning project directly in Model Studio. You will train your models using offline data that are stored in a CAS table. In the **New Project** dialog box, enter a name for the project and select the CAS table that contains the training data.

SAS Visual Data Mining and Machine Learning comes with many prebuilt pipelines that illustrate best-practice strategies for predictive modeling. Since the target variable is an interval target, you can select one of the templates specific for interval targets from the **Template** list. For this example, select **Advanced template for interval target**. Figure 5.3 shows the selections for creating the project.

Figure 5.3: Creating a SAS Visual Data Mining and Machine Learning Project

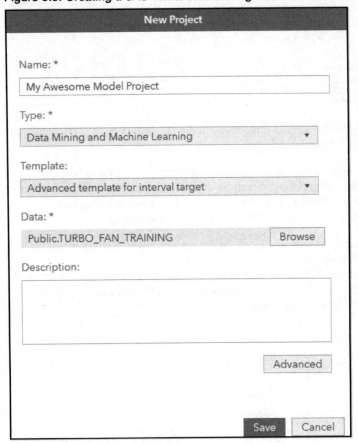

When you create projects from SAS Visual Analytics, the project metadata for the variables in the selected table are automatically set for you according to the roles that you assigned in SAS Visual Analytics. However, when you create the project manually, you must ensure that the desired target variable is specified. You can modify the metadata for the other variables in the project in order to control how those variables are used to train your model.

To predict FuelRatio, select it as the target variable, as shown in Figure 5.4.

Figure 5.4: Defining the Target Variable

	Variable Name	Label	Type	Role	Level		FuelRatio
☐	BypassTotPressure		Numeric	Input	Binary		Role:
☐	CoreSpeed		Numeric	Input	Interval		Target
☐	datetime		Numeric	Rejected	Interval		Level:
☐	Enthalpy		Numeric	Input	Nominal		Interval
☐	FanInletPressure		Numeric	Rejected	Unary		Order:
☑	FuelRatio		Numeric	Target	Interval		
☐	HPTCoolantBleed		Numeric	Input	Interval		Transform:
☐	LPTCoolantBleed		Numeric	Input	Interval		

Click the **Pipelines** tab, and you will see that the pipeline that was selected during the project creation is fully built, has been automatically added to the project, and is ready for use. Figure 5.5 shows the pipeline.

Figure 5.5: Initial Pipeline

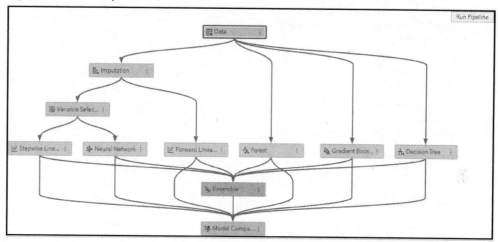

You can modify this pipeline by adding or removing nodes and by modifying the properties of the nodes. You can also add additional pipelines to the project by using any of the other available pipeline templates, or by using the blank template and building a pipeline manually. For more information about all the templates available, see the section "Available Templates" in *SAS Visual Data Mining and Machine Learning: User's Guide* in the SAS documentation.

When you run a pipeline, a champion model is automatically chosen for the pipeline according to predefined model comparison rules. You can modify the model comparison rules by selecting the **Model Compare** node and updating the properties to select the desired comparison options. You can also update the project settings to set the model comparison rules for all pipelines in the project. For more information, see the section "Overview of Model Comparison" in *SAS Visual Data Mining and Machine Learning: Reference Help* in the SAS documentation.

You can see the champion model that was chosen by viewing the results for the Model Compare node. Figure 5.6 shows that the gradient boosting model was chosen as the champion model for this example.

Figure 5.6: Model Comparison Results

Champion	Name	Algorithm Name	Average Squared Error	Root Average Squared Error
☒	Gradient Boosting	Gradient Boosting	0.0161	0.1269
	Decision Tree	Decision Tree	0.0633	0.2517
	Ensemble	Ensemble	0.0666	0.2581
	Forest	Forest	0.0824	0.2871
	Forward Linear Regression	Linear Regression	0.0909	0.3015
	Stepwise Linear Regression	Linear Regression	0.0909	0.3015
	Neural Network	Neural Network	0.2799	0.5291

Although a single champion model is automatically selected for each pipeline, you can add additional models for final comparison by adding them as challenger models. To add challenger models, select the desired modeling node and select **Add challenger model** from the node menu. Figure 5.7 shows how you can add a decision tree as a challenger model for this example.

Figure 5.7: Adding Challenger Models

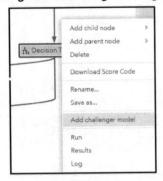

Choosing a Champion Model

The **Pipeline Comparison** tab displays all the candidate models in your project. These models include the champion model for each of the pipelines in the project and any challenger models that you have added. SAS Visual Data Mining and Machine Learning automatically chooses a champion model for your project. You can use this project champion, or you can override it by selecting the desired model and choosing **Set as champion** from the project pipeline menu, as shown in Figure 5.8.

Figure 5.8: The Project Pipeline Menu

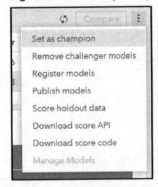

If you want to further validate the project models, then you can select **Score holdout data** from the menu. You are prompted to select a CAS table that contains additional test data that are separate from the data that were used to train the models. New assessment statistics are generated for each of the models, and you can use that information to help in determining the champion model.

For this example, you override the champion model by setting the challenger decision tree model as the champion model.

Deploying the Model to SAS Event Stream Processing

You can deploy the champion model to your SAS Event Stream Processing environment so that you can use the model to score streaming data in real time. The four steps in deploying the model to SAS Event Stream Processing (registering the model, preparing the model, building the project, and deploying the project) are described in the following subsections.

Registering the Model in SAS Model Manager

Registering your SAS Visual Data Mining and Machine Learning models to SAS Model Manager enables you to store them in a common model repository alongside your other analytical models. Storing the model enables you to compare heterogeneous models, run tests against the models, publish models, monitor model performance, and create custom workflows for your business processes.

You can use the common model repository to separate your project and model content, and to set permissions for objects within a repository. For example, you can have different repositories for test and production environments, or for different organizations. The default model repository for SAS Visual Data Mining and Machine Learning is DMRepository.

In the example, register the champion model to SAS Model Manager by first selecting the model in the **Pipeline Comparison** tab. From the project pipeline menu (see Figure 5.8), select **Register models**. Figure 5.9 shows the result of registering the champion decision tree model.

Figure 5.9: Model Registration Status

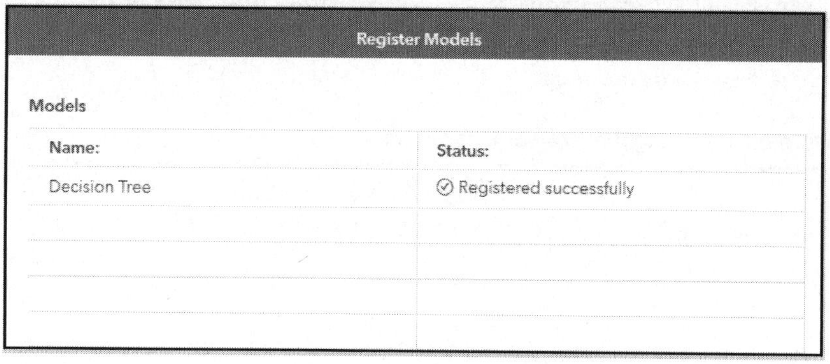

Register Models	
Models	
Name:	Status:
Decision Tree	⊘ Registered successfully

After you have registered the model, a new SAS Model Manager project is created, and the model artifacts (score code, analytics stores, and so on) are associated with the new project. You can navigate to the SAS Model Manager project from SAS Visual Data Mining and Machine Learning by selecting **Manage Models** from the project pipeline menu in the **Pipeline Comparison** tab.

Preparing the Model for SAS Event Stream Processing

Now that the champion model is registered with SAS Model Manager, there are a few steps that you need to take to prepare the model for deployment to SAS Event Stream Processing.

1. Navigate to the **Model** view for the registered model in SAS Model Manager by selecting **Manage Models** from the project pipeline menu. (See Figure 5.8.) Figure 5.10 shows the registered content for your champion model.

    Figure 5.10: Model View

2. Select the **Properties** tab.
3. From the **Target level** list, select **Interval** because the level for the target variable is interval.
4. From the **Output prediction variable** list, select EM_PREDICTION.
5. Save the changes by clicking the **Save** icon ( ) on the toolbar.
6. Navigate to the **Project View** by clicking the project name link.
7. Set the champion model for the SAS Model Manager project by selecting the desired model and choosing **Set as champion** from the menu ( ). Figure 5.11 shows how you set the decision tree model as the champion model.

    Figure 5.11: Setting the Project Champion

Name	Role	Model Function	Project Version	Algorithm	Date Modified	Modifie
Decision Tree (Pipeline 1)		Prediction	Version 1 (1.0)	Decision tree	Mar 15, 2019 01:04 AM	emdusc

Building the SAS Event Stream Processing Project

To deploy your model to your SAS Event Stream Processing environment, you can import the model directly from SAS Model Manager into SAS Event Stream Processing Studio project. For more information, see *SAS Event Stream Processing: Using SAS Event Stream Processing Studio* in the SAS documentation.

From SAS Event Stream Processing Studio, navigate to the **Projects** view and click **New** to create a new project. For this example, name your project FuelRatio.

Defining an Event Source

After you have created the project, you need to define an event source. From the **Windows** view, expand **Input Streams** and drag a **Source** window into the continuous query. Figure 5.12 shows the **Windows** view.

Figure 5.12: SAS Event Stream Processing Studio Windows View

For this example, you simulate the streaming data by configuring the **Source** window to read data from a CSV file. In production, however, your model would get its data from an actual data source. For more information about the supported event sources, see *SAS Event Stream Processing: Connectors and Adapters* in the SAS documentation.

To configure the source to read the CSV file, from the **Source** window properties, expand the **Input Data (Publisher) Connectors** section and add a new entry. The entry has a **Connector type** of **File/Socket Connector**, and you will need to specify the path to the CSV file that contains the data. Figure 5.13 shows the sample source configuration for the example.

Figure 5.13: Defining a Source Connector

Connector Configuration

Name:

New_Connector_1

Connector type: *

File/Socket Connector ▾

☐ Use property values from the file "connectors.config"

☐ Snapshot

Fsname: *

/tmp/tf_all.csv

Fstype: *

csv ▾

All properties...

OK   Cancel

You also need to define the schema for the data ingested by the **Source** window. Figure 5.14 shows the schema that is defined for the example data.

Figure 5.14: Defining the Source Window Schema

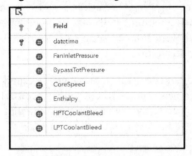

		Field
⚲	⬡	datetime
	⬡	FanInletPressure
	⬡	BypassTotPressure
	⬡	CoreSpeed
	⬡	Enthalpy
	⬡	HPTCoolantBleed
	⬡	LPTCoolantBleed

Importing the Model from SAS Model Manager

After you have defined an event source, you can now import the model. You do this by adding a **Calculate** window to the continuous query and connecting it to the **Source** window.

From the **Windows** view (see Figure 5.12), expand the **Analytics** section and drag a **Calculate** window into the continuous query. Connect the **Source** window to the **Calculate** window by selecting the **Source** window and drawing a connection to the **Calculate** window.

Figure 5.15 shows the continuous query with the Source window connected to the Calculate window.

Figure 5.15: Continuous Query with a Source and Calculate Window

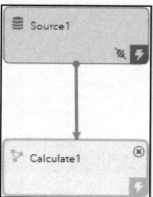

To import the model:

1. Click in the **Calculate** window to display the window properties panel.
2. In the properties pane, expand the **Settings** section, and select **User-specified** from the **Calculation** list. An undefined handler is automatically added to the handlers list.
3. Edit this handler by selecting it and clicking the Edit icon (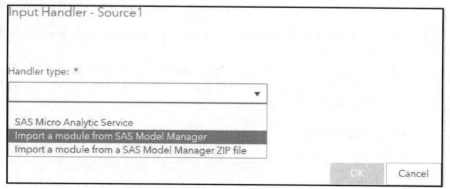).
4. Select Import a module from SAS Model Manager from the Handler type list, as shown in Figure 5.16.

Figure 5.16: Selecting an Input Handler

Input Handler - Source1

Handler type: *

SAS Micro Analytic Service
Import a module from SAS Model Manager
Import a module from a SAS Model Manager ZIP file

OK   Cancel

5. The **Import from SAS Model Manager** dialog box appears, as shown in Figure 5.17. Select the decision tree model that you registered to SAS Model Manager.

Figure 5.17: Selecting the Model to Import

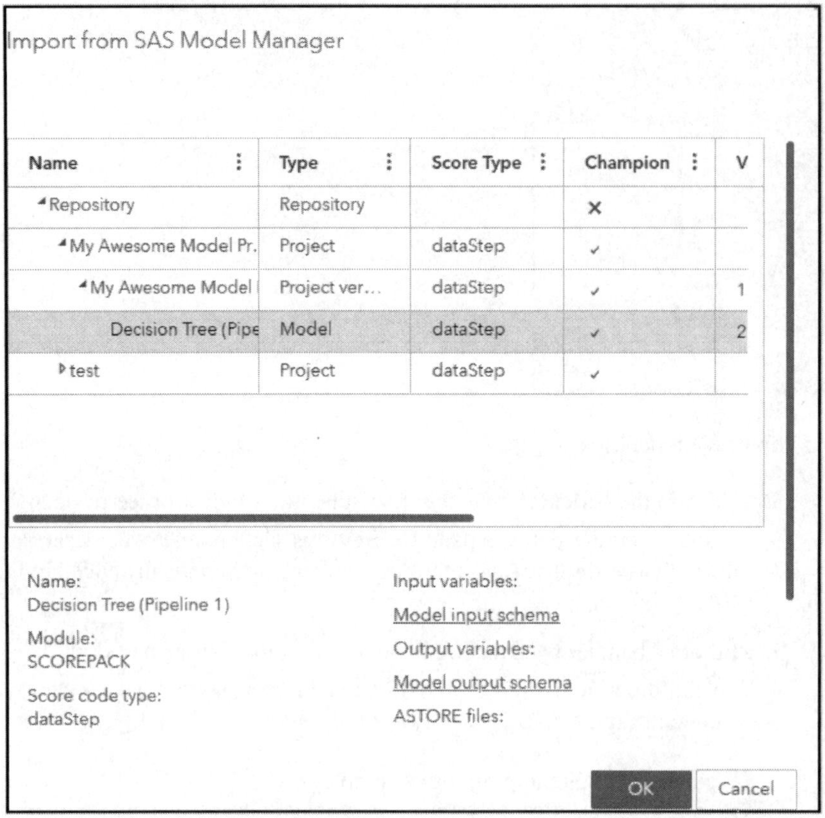

6. Click **OK** to close the dialog box, and then click **OK** again to close the **Input Handler** dialog box.

7. You can configure the **Calculate** window to write the score output to a new file. To do this, expand the **Subscriber Connectors** section and define a new **File/Socket Connector**.

Testing the Project

You can now test the project in SAS Event Stream Processing Studio. You must have a SAS Event Stream Processing server running and registered in SAS Event Stream Processing Studio. For more information about how to do this, see the section "Managing ESP Servers in SAS Event Stream Processing Studio" in *SAS Event Stream Processing: Using SAS Event Stream Processing Studio* in the SAS documentation.

You enter test mode by clicking **Enter Test Mode** in the project toolbar, as shown in Figure 5.18.

Figure 5.18: Project Toolbar

In the **Test** view, click **Run Test** to start the test. Figure 5.19 shows the buttons used to start and stop the test.

Figure 5.19: Test Control Buttons

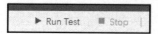

As the test runs, you will see the output for each of the windows that are defined in the project. The **Source** window (see Figure 5.20) shows the events that were processed, and the **Calculate** window (see Figure 5.21) shows the results of scoring each event against the model.

Figure 5.20: Source Window Output

Figure 5.21: Calculate Window Output

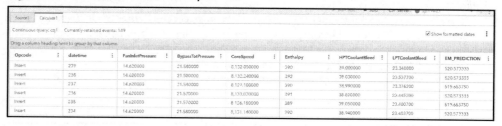

Click **Stop** to end the test. (See Figure 5.19.)

Publishing a New Version of the SAS Event Stream Processing Studio Project

When you are satisfied with your project, click the **Versioning** icon ( ) on the project toolbar (see Figure 5.18).

From the **Versioning** view, select the **Publish a new version** icon ( ), as shown in Figure 5.22.

Figure 5.22: Versioning Toolbar

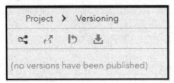

You can add optional notes, and then click **OK** to create the new version. The newly created version appears in the **Versioning** view. Figure 5.23 shows the newly created version of the project.

Figure 5.23: New Version Created

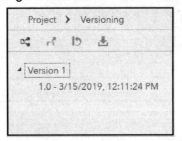

Deploying the SAS Event Stream Processing Project

To deploy the SAS Event Stream Processing Project to SAS Event Stream Processing, you use SAS Event Stream Manager. When you publish a version of a SAS Event Stream Processing Studio project, the published version becomes available for deployment in SAS Event Stream Manager.

If you navigate to SAS Event Stream Manager and select the **Projects** view, then you see your project in the list. Figure 5.24 shows the project that you published in the list.

Figure 5.24: SAS Event Stream Manager Projects View

Name ↑ ⋮	Production ⋮	Tags ⋮	Latest Version ⋮	Published ⋮
FuelRatio	Yes		1.0	3/15/2019, 12:11:24 PM

You can choose whether the project is a production project by right-clicking the project and selecting **Toggle production** from the pop-up menu.

To deploy the project, you need to define a job template. For more information about how to define job templates, refer to the SAS Event Stream Processing documentation.

After you define a job template to deploy the project, navigate to the **Job Templates** view.

Select the desired job template and click **Run a job** using the template icon ( ). Figure 5.25 shows the toolbar options.

Figure 5.25: Job Templates View Toolbar

In this example, a job template called Load Project is defined and is used to deploy the project to the production SAS Event Stream Processing server. Figure 5.26 shows the typical options that are needed to deploy a project. The job requires you to select the desired deployment, project, version, and server.

Figure 5.26: Sample Job Template for Loading Projects

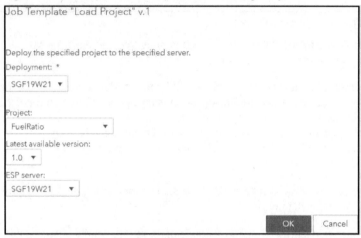

Click **OK** to run your job. When the job completes, you see the job completion status for each of the servers to which you deployed the project. Figure 5.27 shows the completion status for the sample job.

Figure 5.27: Load Project Status

After your project is deployed, you can navigate to the **Deployments** view to see which projects are running on which servers. When you click on a deployment, you will see information about the servers and projects available in that deployment. Figure 5.28 shows the status of a sample deployment with a single server. You can see that the FuelRatio project is running on the server that was specified in the preceding job.

Figure 5.28: Sample Deployment Status

From within a deployment, you can open a running project to view the data that are associated with each window in the project (Source window, Calculate window, and so on).

Monitoring Model Performance

Now that you have deployed the model into your environment, you can use SAS Model Manager to view the performance of the model.

You can collect performance data that have been created by the model at intervals that are determined by your organization. A performance data set is used to assess model prediction accuracy. It includes all the required variables in addition to one or more actual target variables. For example, you might want to create performance data sets monthly or quarterly and then use SAS Model Manager to create a performance definition that includes each time interval.

You can allow SAS Model Manager to score a data set against the model, or you can provide a data set that already contains the predicted values.

SAS Model Manager generates plots – such as variable distribution, characteristic, stability, lift, Gini, ROC, Kolmogorov-Smirnov (KS), and average squared error (ASE) charts – so that you can visualize how the model is performing. Figure 5.29 shows the charts for the example project.

Figure 5.29: Model Performance Charts in Model Manager

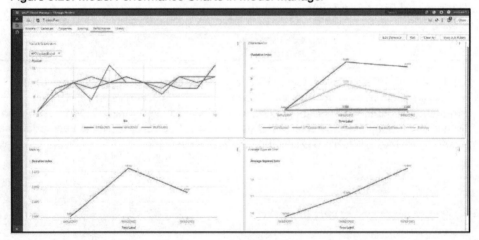

For more information about model performance, see the section "Monitoring Performance" in *SAS Model Manager: User's Guide* in the SAS documentation.

Rebuilding Models

If you notice that the performance of your SAS Visual Data Mining and Machine Learning model has degraded, then you can retrain the registered model from SAS Model Manager.

To retrain a model from SAS Model Manager, navigate to the **Project** view for the project that contains the model and select **Retrain** from the menu ( ) at the top of the **Project** view. The **Retrain Project** dialog box appears, as shown in Figure 5.30.

Figure 5.30: Retraining a Model from Model Manager

If you select **Set the project retrain state to needed**, then you can open the SAS Visual Data Mining and Machine Learning project and retrain your models using new data. Before you run the SAS Visual Data Mining and Machine Learning pipelines, you have the option of modifying the settings of existing modeling components and adding new ones. You can then run the pipelines in the project to recalculate a champion model for the project. You can choose which models will need to be registered or published (or both) after the retraining operation completes.

If you select **Retrain now with a new data source**, then you are prompted to select a CAS table that contains the new data. The SAS Visual Data Mining and Machine Learning project is automatically retrained with the selected data. The project pipelines are run automatically, and the champion model for the project is registered in SAS Model Manager.

For this example, select **Set the project retrain state to needed**. Then navigate back to your SAS Visual Data Mining and Machine Learning project in Model Studio. After you are in your SAS Visual Data Mining and Machine Learning project, from the **Data** tab, expand the **Data sources** view and click the **Replace data source** icon ( ). Figure 5.31 shows the project data sources view.

Figure 5.31: Project Data Sources View

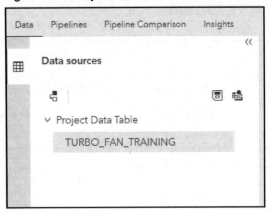

Select a new CAS table that contains the updated data that you want to use to retrain your models. Once you have selected a new table, you need to rerun any pipelines in your project. However, before you run the pipelines, you can modify them as needed.

When you run your pipelines, champion models are recalculated for each pipeline. All the champion models and any previously defined challenger models are shown in the **Pipeline Comparison** tab.

In the example, you previously had overridden the calculated champion model to select the decision tree as the project champion. This setting was remembered, and the decision tree is once again set as the project champion.

However, as you look at the results, you might determine that a different model performed better against the new data. In this example, you add the forward linear regression model as a challenger model and select it as the new project champion.

You register this new model to SAS Model Manager and navigate to the model project in SAS Model Manager. Select the newly registered linear regression model and mark it as the new project champion. Figure 5.32 shows the new champion model selection.

Figure 5.32: Selecting a New Project Champion in SAS Model Manager

Deploying the Updated Model

When you navigate back to SAS Event Stream Manager, you will notice a notification that an update is available for your FuelRatio project. Figure 5.33 shows the update notification.

Figure 5.33: SAS Event Stream Manager Notification

Select **Deploy**, and you have the option to update all projects or update only selected projects, as shown in Figure 5.34.

Figure 5.34: Deploying Updates

For this example, select **All projects running this version** and click **OK**.

When you navigate to the **Projects** view, you will see that your project version has been incremented, and the **Deployments** view shows that the new version is now running in your environment.

Now you can repeat the process of monitoring the newly deployed model and retraining as needed.

Conclusion

An IoT reference architecture provides the backbone of a robust, secure, and extensible IoT solution. This architecture is not just a collection of tools; it is a finely tuned ecosystem with the important characteristics described in this chapter. Such an architecture must meet a set of challenges in order to provide business value to customers. SAS Event Stream Processing, an important part of an IoT reference architecture, can be used to meet those challenges. It readily provides the multi-phase analytics that you can use to efficiently extract intelligence from the edge and can be used in the cloud to simplify the deployment and management of IoT solutions. This versatility enables it to keep up with the rapid growth of "things" in the IoT universe.

Other SAS products serve important roles in the IoT reference architecture. You explore and visualize any data with SAS Visual Analytics. You can examine and understand patterns, trends, and the relationships in your data and apply various predictive analytics models and visualizations to begin gaining valuable insights. You can then take the models that you have created in SAS Visual Analytics and create analytics pipelines in SAS Visual Data Mining and Machine Learning. You can augment these pipelines with additional models, continually tuning and comparing them until you have built the best model for your data. You can deploy your model from SAS Visual Data Mining and Machine Learning by publishing the model directly to a location that contains the data that you want to score. You can also register the model in SAS Model Manager to take advantage of model versioning and governance. From SAS Model Manager, you can deploy your model into your SAS Event Stream Processing environment to score streaming data at various levels, including IoT gateways and edge devices.

As your data evolve, you can use SAS Model Manager to monitor your model's performance over time so that you can make decisions about whether you should rebuild your models. If you decide to rebuild your model, you go back to the beginning of the analytics life cycle. All these steps are contained within the IoT reference architecture.

References

SAS. SAS Event Stream Processing 6.2 Documentation. https://support.sas.com/en/software/event-stream-processing-support.html#documentation.

Saxena, A., and Goebel, K. (2008). "Turbofan Engine Degradation Simulation Data Set." Accessed January 17, 2017. NASA Ames Prognostics Data Repository. NASA Ames Research Center, Moffett Field, CA. https://ti.arc.nasa.gov/tech/dash/groups/pcoe/prognostic-data-repository/.

About the Contributors

As the Principal Product Manager for SAS Event Stream Processing at SAS, **Stephen Sparano** focuses on Internet of Things (IoT) platform technologies to develop the SAS ESP vision and strategy and help guide research and development teams at SAS to develop high performance, in-memory streaming analytics software and solutions. Stephen holds a bachelor's degree from Providence College.

As a Principal Software Developer in Advanced Analytics, **Shawn Pecze** is responsible for developing and supporting the services that comprise SAS Enterprise Miner, SAS Factory Miner, and SAS Visual Data Mining and Machine Learning. Shawn holds a master's degree from Rensselaer Polytechnic Institute and a bachelor's degree from Penn State University.

As a Senior Software Developer in the Advanced Analytics Division at SAS, **Prasanth Kanakadandi** works on the design and development of innovative, high quality and scalable software solutions. Prasanth holds a master's degree in Computer Science from New Mexico State University and a bachelor's of technology degree in Computer Science and Engineering from Jawaharlal Nehru Technological University, India.

Byron Biggs is a Principal Research Statistician in the Internet of Things Division at SAS, where he specializes in analytics for high-frequency streaming data and high-performance computing. Prior to SAS, Byron worked in high-performance distributed computing for software companies and in physics/software consulting. Byron holds a PhD in Physics from the University of Virginia and a BSE in Electrical Engineering and Computer Science from Princeton University.

Chapter 6: Artificial Intelligence and the Internet of Things

By Chip Robie and Michael Harvey

Introduction

In the living room the voice-clock sang, Tick-tock, seven o'clock, time to get up, time to get up, seven o'clock! …

"Today is August 4, 2026," said a second voice from the kitchen ceiling, "in the city of Allendale, California." It repeated the date three times for memory's sake. "Today is Mr. Featherstone's birthday. Today is the anniversary of Tilita's marriage. Insurance is payable, as are the water, gas, and light bills." …

It was raining outside. The weather box on the front door sang quietly:
"Rain, rain, go away; umbrellas, raincoats for today..." ...

Nine-fifteen, sang the clock, time to clean.

Out of warrens in the wall, tiny robot mice darted. The rooms were a crawl
with the small cleaning robots, all rubber and metal. They thudded against
chairs, whirling their moustached runners, kneading the rug nap, sucking
gently at hidden dust. Then, like mysterious invaders, they popped into their
burrows. Their pink electric eyes faded. The house was clean.

"There Will Come Soft Rains," Ray Bradbury (1950).

We are fascinated with the future. We like to speculate how current technology might evolve
during the coming decades or centuries. The smart automated home that Ray Bradbury
described in 1950 included digital assistants with natural language processing and
synthesized speech, smart kitchen appliances, vacuuming robots, programmable flat-screen
display walls, and even smart garage and home doors that used schedules, motion sensors,
and facial recognition to provide keyless access and security. Seventy years later, many of
these fantastical notions are real consumer goods that are widely available in an
interconnected world economy.

It's just as fascinating to consider how a period's emerging technologies influenced its vision
of the future. For example, during the 1950s in the US, the future was filled with rockets and
robots. Contemporary entertainment reflected that theme. In 1956, MGM's "Forbidden
Planet" introduced an artificially intelligent mechanical servant named "Robby the Robot."
The robotic theme continued into the mid-1960s as television shows such as "Lost in Space"
depicted family-based interplanetary travel adventures helped by the General Utility Non-
Theorizing Robot who could pilot space craft, detect danger, and play the guitar.

As the 1960s progressed, more futuristic gadgets appeared as we ventured where no one had
gone before. Star Trek (and its successors) introduced us to the concepts of wireless
communication, digital medical diagnostics, global location tracking, and sophisticated
android technology. After Neil Armstrong and Buzz Aldrin set foot on the moon in 1969, our
futuristic fantasies soared into overdrive. In the 1970s, Star Wars brought us worlds with
embedded devices, as well as countless "droids" and independent digital intelligence entities.

More than 40 years after the first Star Wars film, many of those futuristic dreams have
become reality. The growth of the deep learning and machine learning fields has fueled the
pursuit of the practical applications of artificial intelligence (AI). The broad availability of
cloud computing with its attendant advantages in data storage and computation speeds,
combined with growing trends in wireless networking and the Internet of Things have rapidly
accelerated this pursuit. The results emerge in the many new AI-inspired innovations that are
surfacing in the contemporary world, elevating cultural norms about how we deploy and use
smart and embedded technology. Today's devices that use real machine learning and artificial
intelligence are leapfrogging past our earlier visions of the future. Most of those devices are
smarter than the computers that helped Armstrong and Aldrin land on the moon.

What Do We Mean by Artificial Intelligence?

Artificial intelligence dominates current information technology industries as much as rockets and robots dominated 1950s and 1960s popular culture. AI emphasizes the creation of intelligent machines that use structured deep networks and self-optimizing training algorithms (instead of explicitly coded instructions) when learning how to perform a task. The task can be as simple as classifying and labeling an object in an image, or as complex as interpreting hundreds of items detected by a sensor array in real time. Fulfillment of the task can be something as mundane as identifying the emotion on a face or something as complicated as operating a vehicle without human intervention.

There are two types of AI: weak and strong. *Weak AI* is artificial intelligence focused on a single task. The machine is built in a way that it can independently perform a task that makes it seem smart, such as playing poker, after all rules and potential moves have been fed to the machine. All possible scenarios are entered to train the machine. Weak AI systems can eventually contribute to building strong AI.

Strong AI results in a machine that can think and perform tasks on its own like a human being. There are not yet any examples of strong AI currently in use, but research is making exciting progress. Unless specifically noted otherwise, this chapter refers to weak AI when using the term *AI*.

Typical examples of AI include the following:

- Machine learning includes supervised and unsupervised deep neural network learning.
- Natural Language Processing (NLP) includes content extraction, language classification and machine translation, question answering, and text generation.
- Expert systems for speech involve the transformation of speech to text, and text to speech.
- Expert systems for vision include machine vision, object detection, image classification, and image recognition tasks.
- Further expert systems exist for AI planning, logistics, and robotics.

Machine learning uses a deep neural network with many input variables and a defined target. The deep neural network learns how to approach target values using an iterative series of algorithmic training steps that "teach" the machine correct task resolutions. One instance of machine learning is an image classification program. Such a program learns how to identify different classes of objects not previously seen through training the machine with labeled classes of similar objects. For example, after a machine is trained with labeled images of pets (cat, dog, rabbit), it can recognize and classify images of cats, dogs, and rabbits that were not originally included as part of the model training data.

Tasks such as object identification and image classification are performed by deep CNN networks. The following diagram represents the functional layers that comprise a deep learning convolutional neural network (Figure 6.1).

Figure 6.1: Typical Deep Convolutional Neural Network (CNN) Architecture for Image Classification Tasks

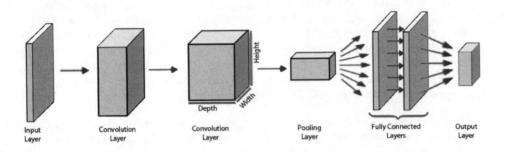

Machine vision systems capture and analyze visual information. Data from cameras that produce digital image streams are analyzed for a variety of applications:

- High-speed manufacturing and quality control programs
- Home security applications such as those offered by Ring
- Live chat apps such as FaceTime
- Crowd security systems at municipal and entertainment venues
- Various levels of automatic vehicle driving
- Image analysis diagnostics

Natural Language Processing is manipulation of natural language by software to render a machine understanding of the concepts embedded in the spoken words or text. Common NLP examples today are chat-bots, international social media comment translators, as well as real-time digital assistants such as Google's Echo and Amazon's Alexa. The latter two applications can both understand and respond to verbal requests in multiple languages.

Speech synthesis systems are machine learning artifacts that enable machines to parse and interpret written or spoken language and perform an associated task, such as analyze the sentiment in a speech segment and return a written or spoken response or action based on the submitted content. Text to speech examples abound. The Siri assistant from Apple and the Waze interactive navigation system from Google are probably the most widely known.

Intersections between machine learning, artificial intelligence, and the Internet of Things are rapidly evolving. Examples abound. Social media platforms such as Facebook and Instagram burst with embedded algorithms that recognize familiar faces in pictures and cross-reference recent internet browser histories with advertising opportunities. Relevant advertisements then surface in social media feeds, email, electronic coupons, and then appear on e-shopping vendor sites such as Amazon and eBay.

As we increasingly adapt to and integrate smart technology in the forms of phones, tablets, watches, door locks, security cameras, geo-fences, GPS-based auto navigation apps, diagnostic medical devices, and so on, an enormous amount of metadata is generated. As we encapsulate and interpret the information contained in that metadata, we better understand how to pool, store, and use it to perform more intelligent, computer-aided tasks.

The generation and integration of smart devices, their associated data streams, and the digital information exchanges enabled by wireless networks, WiFi transmitters, and 5G data sharing will shape our immediate technological future.

How Does AI Interact with the Internet of Things?

What are some of the emerging trends and important changes happening related to AI and the Internet of Things (IoT) that we should understand and prepare for? The sections that follow attempt to answer that question.

Increasing Numbers of Smart Connected Devices

In November 2018, the Gartner global research and advisory firm addressed the top Internet of Things (IoT) technology trends that are expected to drive digital business innovation over the next five years. The first and strongest IoT trend identified was the inclusion of devices that use artificial intelligence algorithms to interact with users and the digital ecosystem of interconnected high-speed, wireless, cloud-based, and 5G networks.

Figure 6.2 shows SAS Event Stream Processing (SAS ESP) implemented with an array of connected sensors. For more information about processing real-time operations using SAS Event Stream Processing, see "Want real-time predictive analysis? You need Event Stream Processing" by David Pope.

Figure 6.2: Batch and Real-Time Operations in an Industrial Security System using SAS ESP

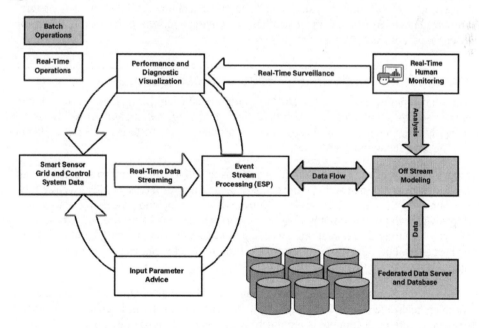

Gartner forecasted that 14.2 billion smart connected devices would be in use during 2019, reaching 25 billion devices by 2021. These devices will produce immense amounts of metadata that can be researched, repurposed, and reused to train a wide variety of model and expert systems. Nick Jones, research vice president at Gartner, states that, "Data is the fuel that powers the IoT... AI will be applied to a wide range of IoT information, including video, still images, speech, network traffic activity, and sensor data" (Gartner 2018).

New Infonomics of Accumulated Smart Device Metadata

According to Gartner, *infonomics* is "the theory, study and discipline of asserting economic significance to information. It strives to apply both economic and asset management principles and practices to the valuation, handling and deployment of information assets" (Laney 2018).

IoT exists because the data generated by devices in the world have significant value to businesses. The proliferation of smart devices comes with a comparable growth in device metadata. Smart refrigerators, thermostats, door locks, security cameras, watches, tablets, phones, and digital assistants collect all types of data – when you are home, when you are at work, preferred home temperatures, waking and sleeping hours, shopping preferences and frequencies, embedded financial and credit information, GPS locations, vehicle speed and acceleration parameters, preferred commuting routes, dining out frequency, phone and spending records, browser searches, and much more.

With the exponential growth in generated and digitally stored data, infonomics and data brokering has become a technical business opportunity. In their 2018 survey, Gartner found 35% of respondents were selling or planning to sell data that was collected by their products and services. Monetizing the value of collected metadata through infonomics is a strategic growth strategy. By 2023, Gartner predicts that buying and selling IoT data will become an essential part of the IoT ecosystem.

Moving from Traditional to Edge to Mesh Computing

Current computing trends show a shift from traditional centralized client/server computing enterprises toward cloud and edge architectures. Evolving changes and growth in wireless network infrastructures as well as wireless smart devices are fostering new data collection, sharing, and deployment strategies.

Edge architectures use cloud computing systems to deliver an optimized and distributed analytic approach. Edge architectures deploy computing resources physically close to where the data is being collected, removing recurrent data fetching and processing tasks while improving scalability and optimizing network efficiency. SAS ESP uses edge processing to analyze streaming data in real time, weeding out data noise and passing high quality data to be processed. For more in-depth information about SAS Event Streaming, see https://www.sas.com/en_us/software/event-stream-processing.html.

However, the shift from central client/server to cloud to edge computing is not the end point for many real-time AI and IoT architectural designs. The neatly ordered set of layers that are associated with edge computing architectures are likely to morph into a more unstructured networking architecture created by a wide range of interconnected IoT devices, likely assisted by wireless 5G infrastructures. The collection of interconnected things and services becomes associated together in a "dynamic mesh" (Figure 6.3).

Figure 6.3: Mesh Network Topology

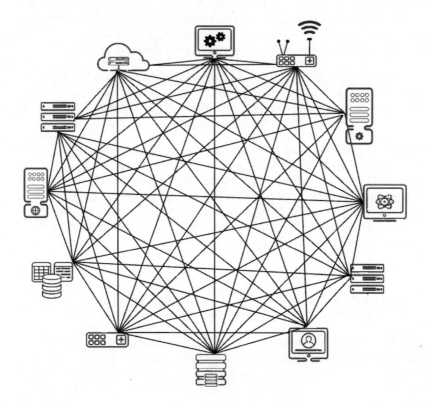

Mesh networks are local networks whose nodes connect directly, dynamically, and without hierarchy to as many other nodes as possible. Nodes in a mesh network cooperate to efficiently route data dynamically: if a specific node in a communication pathway drops out, another node in the mesh takes its place without incident. The resilience and self-organizing nature of mesh networks enable more flexible, responsive, and smart IoT systems. Network architecture experts should prepare for the impact of mesh architectures on growing enterprise IT infrastructure and management resources.

Applications: Integrating AI Technologies with IoT

What are some examples of AI technologies that have merged with IoT to deliver integrated solutions?

As large urban densities increase, the resources and energy necessary to sustain the population must also increase in kind. Cities worldwide now use integrated AI and IoT technologies to interact with traffic control systems and law enforcement to improve urban security and emergency services. A smart city can take advantage of information to enhance the quality and availability of energy services, public transportation, waste management, smart manufacturing, smart buildings, and smart power distribution grids.

Global health-care systems are implementing deep learning object detection and image analysis tools that enhance diagnostic accuracy rates while interpreting X-rays, MRI scans, ultrasound scans, and cytological analyses. Hospital instrumentation can now use wireless data networks, making the data available to practitioners more rapidly and enhancing the ways the data can be analyzed, shared, or repurposed to assist patients in similar conditions.

Industrial manufacturing, logistics, and material handling equipment is being modernized with sophisticated arrays of embedded sensors. Sensor arrays measure operational parameters such as machine speeds, chamber pressures, heater temperatures, material flow rates, QA monitoring and rejection, package registration, and so on.

Manufacturers use edge and mesh networks not only to execute material and product flows necessary for high-speed manufacturing and quality assurance, but also to collect data from arrays of rotation and vibration sensors whose inputs can be used to predict impending component failures and formulate predictive maintenance frequencies and intervals for component and subassembly rebuilds. Integrated manufacturing networks are moving past client/server SCADA (Supervisory Control and Data Acquisition) architectures to embrace edge and mesh device computing supported by machine learning algorithms.

Consumer fashion has embraced wearable smart devices, speeding their integration in the marketplace. Smart watches can handle electronic communications and smart apps. They can also work together with sensors that monitor physiologic data, such as pulse, workout duration, blood glucose levels, and body temperature.

Industrial wearables, such as glove bar code readers and wrist-mounted warehouse computers, use wireless networks and repeaters to expedite and audit commercial order fulfillment transactions. Soldiers in the field are replacing tactical handheld tablets and smartphones with wireless wearable computers that are loaded with tactical, geographic, and operational data. The wearable computers enable remote team monitors as well as on-mission team members to track and communicate with team formations, lending real-time guidance and communication in the harshest environments.

Hikers, international adventurers, and even patients with dementia can wear clothing with embedded GPS tracking beacons. These beacons generate virtual map trails, waypoints, and determining emergency rescue locations. Children's watches such as the Eoncore GPS/GSM/Wifi Watch for Kids now enable adults to verify the location of a child on demand. The Garmin Vivofit JR watch for kids includes a Chore Reminder app that interacts with parents to monitor and reward kids for completed chores, such as washing dishes or putting away laundry. Fitness watches track activities while monitoring heart rates and can even monitor sleep cycles.

Law enforcement and the military are integrating data from wearable devices that provide audio and video recordings of events, time stamped and indexed by GPS location. Small wearable personal safety devices worn on belt-loops (such as Instinct by Revolar) use geofencing to notify pre-selected individuals when the wearer is safely at home (or work), or in transit. Click settings allow yellow and red alerts, which request help from the listed individuals or 911 Emergency.

Digital twinning is now possible through IoT and wireless networking. A digital twin is a virtual model of a product, service, or process. Digital twinning is a model process that bridges the physical and virtual realms.

In comparison to physical manufacturing equipment that is embedded with an array of sensors and detectors, a digital twin can be created when data collected by physical sensor-embedded smart components is processed by a cloud-based system. The system receives and processes the collected data with other business and contextual data to form a virtual representation of the physical machine. In the virtual environment, time can be accelerated, multivariate relationships can be probed, lessons can be learned, and opportunities can be discovered to transform business practices.

According to NASA's leading manufacturing expert, John Vickers, "The ultimate vision for the digital twin is to create, test, and build our equipment in a virtual environment. Only when we get it to where it performs to our requirements do we physically manufacture it. We then want that physical build to tie back to its digital twin through sensors, so that the digital twin contains all the information we could have by inspecting the physical build" (Marr 2017). As more companies learn about the successes gained by digital twinning, they will in turn want to deploy their own digital twins to gain competitive advantages. You can read about the application of a digital twin in Chapter 8.

There's No Place Like Home: AI and IoT

As pointed out at the beginning of this chapter, more than 70 years ago, scientists and fantasy writers such as Ray Bradbury predicted smart homes. In a 2018 survey, Cisco found that connected home applications such as home automation, video surveillance, home security, connected white goods, and tracking applications will represent nearly half (46%) of the total machine-to-machine network connections by 2021 (Figure 6.4). That's immense! With that type of momentum, what is the current reality of AI and IoT usage in consumer homes?

Figure 6.4: Integrating Home Automation with IoT

An arrival notice event signals your Nest smart home thermostat, which resets the HVAC from energy-saving mode to a cool 70-degree F setting. As you walk to your door, your Ring door camera uses an AI facial recognition algorithm to identify you, then opens the smart-connected August door lock. As you enter the house, a sequence of events is triggered: Google Home turns on the den and kitchen lights. Alexa reminds you that your favorite streaming network shows is on tonight, then asks whether you would like to record it. Afterward, Alexa reads you your preferred news headlines and offers to play your favorite streaming internet music station (or talk radio) while you prepare dinner.

If you wonder where the nearest nail salon is, or if you want to order delivery food, you merely need to verbally ask a digital assistant such as Siri, Alexa, or Google Echo. Voice activation is a huge gain for everyone, especially the impaired—being able to turn on lights in any part of the house, or place and receive phone calls, or change thermostat settings while remaining in a chair or bed is an enabling and welcome lifestyle improvement.

If your smart home has Ring, Nest, or SimpliSafe sensors inside, while you are at work, you might be notified of a loud noise in your kitchen. From your app, you might view the kitchen camera and discover that your pet Fluffy has spilled a bowl of dry cat food on the floor. Bad Fluffy! Not wishing to come home to a mess, you use an app on your smart phone to activate your Wi-Fi enabled robot vacuum cleaner and instruct it to clean up the home kitchen zone… from the chair of your office at work.

The new conveniences we gain from the merging of the IoT and smart technology are great. However, as deployments move from researchers to commercial customers, installing and using new data-driven conveniences comes with the expense of a learning curve. Even technologies like email have important rules for use and security, which, if ignored, can lead to potential disruptions by hackers or bad actors.

Even so, the future remains bright…even if we are not wearing the Google glasses that might have been a bit too much of a future shift for many of us.

Creating and Remotely Deploying a SAS Deep Learning Image Detection and Classification Model

Earlier in the chapter, we described typical AI tasks such as classifying and labeling objects in an image. Here, we provide an example of SAS Deep Learning code that could be used to perform this task. What follows are the steps required to create a CNN (Convolutional Neural Network) image detection and classification model. The model is built using the SAS Viya Deep Learning tools and the SAS DLPy Python application programming interface (API). SAS DLPy is a high-level Python library for the SAS Deep Learning features in SAS Viya. After the model has been created, validated, and stored in transportable ASTORE format, the ASTORE model is deployed in a separate SAS Event Stream Processing environment to score a stream of incoming images.

Rather than build a model from scratch, this example uses DLPy libraries to import a pre-built open-source VGG16 model. It also imports a set of pre-trained weights for the model. Starting a modeling project with a validated architecture and initial training weights can save a lot of time. The pre-built, open-source VGG16 model is a proven, 16-layer open-source network used by the VGG team in the 2014 ImageNet Large Scale Visual Recognition Challenge. The classification task analyzes input images of retail product offerings (associated with one of 12 superheroes) and predicts which superhero target class each image represents.

After the image detection and classification model is created, trained, and validated, it is saved to a SAS Analytic Store (ASTORE) format. Using an ASTORE file enables it to be deployed in other environments to score new data. Then, the ASTORE file is used with SAS Event Stream Processing to score new image data.

The training and test data used was part of a 2018 Crowd Analytix data mining challenge (https://www.crowdanalytix.com/contests/identifying-superheroes-from-product-images). The contest data was scraped from SuperHeroDb and released under CCO as public domain. Advanced readers can examine the training and test data used at http://support.sas.com/documentation/prod-p/vdmml/zip/superhero_image_data.zip.

It should be emphasized that this is not ready-to-run code. This example is intended as a demonstration of the developmental steps and code components required to create an image detection and classification model. It simulates the cells of a Jupyter Notebook. Using the appropriate tools and packages, you can use this example to develop ready-to-run code.

The example follows these steps:

1. Initialize libraries and launch SAS CAS
2. Load and explore the training data
3. Prepare the training data for modeling
4. Specify pre-defined model architecture and import pre-trained weights
5. Train the model
6. Score the test data to validate model accuracy
7. Browse the scored data
8. Save model as ASTORE for deployment
9. Upload the ASTORE to CAS
10. Use SAS ESPPy to deploy ASTORE model and score streaming data

Initialize Python Libraries and Launch SAS CAS

The first step prepares your computing environment. You load libraries from the SAS DLPy API and from common Python analytic calculation and plotting libraries such as matplotlib. The SAS Scripting Wrapper for Analytics Transfer (SWAT) package is the Python client to SAS Cloud Analytic Services (CAS). It enables users to execute SAS CAS actions and process the results using the Python API instead of the SAS CAS programming language.

Note: For more information about SAS DLPy, see the SAS DLPy repository on GitHub at https://github.com/sassoftware/python-dlpy.

For more information about starting a SAS CAS session with the SWAT package, see https://sassoftware.github.io/python-swat/getting-started.html.

Example Code 6.1

```
In [1]:
# Import Required Libraries and Utilities
# SAS Statistical Wrapper for Analytics Transfer
import swat
# SAS DLPy Python API Module
import dlpy
# Various SAS DLPy Functional Libraries
from dlpy.images import ImageTable
from dlpy.splitting import two_way_split
from dlpy.applications import *
from dlpy.utils import add_caslib
# Display Matplotlib Graphic Output in Jupyter Notebook Cell
import matplotlib
%matplotlib inline
```

After you import the required libraries and utilities, connect to a SAS CAS session. You must know the name of your CAS server and port ID, as well as your user ID and password.

Example Code 6.2

```
In [2]:
# Create a CAS Connection
conn = swat.CAS(CAS-hostname, port-number, userID, password)
```

Next, specify a path for your model project data. The path lets SAS CAS know where to find your model train and test data, as well as the pre-defined VGG16 model architecture and VGG16 model weights file. Because this model classifies superheroes, the following project path is suggested:

Example Code 6.3

```
In [3]:
# Set Directory Path for Project
dir_path = '/data/userID/Superhero_Classification/'
```

Load and Explore the Training Data

The second step assumes that you have downloaded the model training data CAX_Superhero_Train into the model project directory /data/userID/Superhero_Classification_Model/CAX_Superhero_Train. The code uses the DLPy load_files command to create a CAS table named superTrain, and then loads the superTrain table with the training data in CAX_Superhero_Train.

Example Code 6.4

```
In [4]:
superTrain = ImageTable.load_files(conn,
                                   path=dir_path +
'CAX_Superhero_Train',
                                   casout=dict(name='superTrain')
                                   )
                          )
```

The SAS CAS table superTrain is created and loaded with the model training data. It is a collection of observations that contain retail product images featuring content from one of 12 superheroes.

To get an idea of what the training data looks like, you can use the show() command. Here we display a table of ten randomly selected images from the table superTrain, organized in five columns.

Example Code 6.5

```
In [5]:
superTrain.show(# Show 10 images
               nimages=10,
               # Display the images in 5 columns
               ncol=5,
               # Choose the image observations
               # to be displayed at random
               randomize=True
               )
```

The preview table displays ten random retail product images from the table superTrain. You can see a variety of items: shirts, hoodies, backpacks, bags, even cufflinks, all with a superhero theme.

How are the retail images divided up among the superheroes? How many images do we have for each superhero? You can examine the distribution of superhero classes in the model training data using the label_freq() function with the superTrain table. Here, you view the twelve superTrain superhero classes, along with the frequency of each class in the model training data.

Example Code 6.6

```
In [6]:
superTrain.label_freq
```

```
Out[6]:
```
Frequency for SUPERTRAIN

	Level	Frequency
Ant-Man	1	241
Aquaman	2	201
Avengers	3	216
Batman	4	779
Black Panther	5	460
Captain America	6	410
Catwoman	7	200
Ghost Rider	8	200
Hulk	9	414
Iron Man	10	694
Spiderman	11	861
Superman	12	757

All 12 classes of superheroes appear in the table. The number of images varies among superhero classes; they range from 201 images for Aquaman to 861 images of Spiderman.

You can use the image_summary command to generate a summary table of the attributes of the images in the model data. What is the total number of training images? What about image sizes – what is the range of image dimensions in the model training data? Consistent image size is a positive factor to improve the results of model training.

Example Code 6.7

```
In [7]:
superTrain.image_summary

Out[7]:
jpg                5433
minWidth             73
maxWidth            540
minHeight           120
maxHeight           522
meanWidth       197.982
meanHeight      230.795
mean1stChannel  167.516
min1stChannel         0
max1stChannel       255
mean2ndChannel   166.33
min2ndChannel         0
max2ndChannel       255
mean3rdChannel  173.753
min3rdChannel         0
max3rdChannel       255
dtype: object
```

The training data attribute table contains significant information. There are 5,433 input images in a variety of sizes. The training image heights range from 120 to 522 pixels tall, and image widths range from 73 to 540 pixels wide. However, standard image sizes provide better model training results.

In the following steps, you standardize the image sizes to 224 x 224 pixels. You also use patches from existing training images to create new, similar training images. Augmenting the number of images in training data also results in improved model training.

One other technique to improve model training results is to begin with well-randomized training data. Clusters of similar training images in the data can degrade model training performance. Also note that the mean channel values (for B, G, and R colors) are displayed in the frequency table. The mean channel values are specified again during model training.

Prepare the Training Data for Modeling

With the information in the training data attribute table, you can take steps to improve the model training performance in the third step.

First, randomize the training data by using the shuffle() command with the superTrain table. Using randomized training data is a best practice for model training performance.

Example Code 6.8

```
In [8]:
# Shuffle Images
superTrain.shuffle(casout=dict(name='superTrain',
                               replace=True
                               )
                  );
```

Next, partition the data into train and test partitions using the two_way_split() function with the superTrain table. The function divides the big data set into train and test partitions. A random seed value of 12345 is used for repeatability. The data is partitioned so that 80% of observations are assigned to trainData, and 20% of observations are allocated to the test partition testData.

Example Code 6.9

```
In [9]:
trainData, testData = two_way_split(# table to be split
                                    superTrain,
                                    # 20% to testData table
                                    # Remainder 80% goes to trainData
table
                                    test_rate=20,
                                    # Random seed ID for repeatability
                                    seed=12345
                                    )
```

Now you can improve training data performance by augmenting the training data using portions, or partial "patches" of an existing image to create a new, similar training image. Use the as_patches() function with the trainData table to create new images.

The as_patches() function is configured to create 200 x 200 pixel image patches that have a step_size of 24. In keeping with the plan to standardize all training images to 224 x 224 pixels, this configuration results in new output image patches sized 224 x 224 pixels. Setting inplace=True replaces the original training data with the newly augmented and resized training data. Augmented training data sets are a tip for improved model training performance.

Example Code 6.10

```
In [10]:
trainData.as_patches(# Width of image patch in pixels
                     width=200,
                     # Height of image patch in pixels
                     height=200,
                     # Step size in pixels of patch image
                     step_size=24,
                     # Net image width after as_patches()
                     output_width=224,
                     # Net image height after as_patches()
                     output_height=224,
                     # Overwrite old images with resized images
                     inplace=True
                     )
```

Now, all newly created patch images are uniformly 224 x 224 pixels in size. Action on these patch images will not affect any of the existing training data images that are different sizes. Use the resize() function with the testData table to resize images.

Example Code 6.11

```
In [11]:
# Resize Test Data
testData.resize(width=224,
                height=224,
                inplace=True
                )
```

After standardizing images in the testData partition, verify that all of them (original and augmented) are 224 x 224 pixels in size. Use the image_summary command to display a summary table of the image attributes in the augmented trainData table.

Example Code 6.12

```
In [12]:
trainData.image_summary

Out[12]:
jpg                 79463
minWidth              224
maxWidth              224
minHeight             224
maxHeight             224
meanWidth             224
meanHeight            224
mean1stChannel    157.613
min1stChannel           0
max1stChannel         255
mean2ndChannel     154.45
min2ndChannel           0
max2ndChannel         255
mean3rdChannel    163.318
min3rdChannel           0
max3rdChannel         255
dtype: object
```

The updated table summary confirms that table trainData has been augmented and standardized. The table has increased from 5,433 images to 79,463 images. All images in trainData are standardized to 224 x 224 pixels.

Test the uniform sizing by examining a sampling of the images. Output results should be the same size. Use the show() function with the table trainData to display a table of ten random labeled images using five columns.

Example Code 6.13

```
In [13]:
trainData.show(nimages=10,
               ncol=5,
               randomize=True
               )
```

Specify Predefined Model Architecture and Import Pre-Trained Model Weights

The VGG16 model template VGG16_notop has been saved to a location within the specified project path. The VGG16 pre-trained model weights (in .h5 file format) are also saved in that location. The following Python code uses the SAS DLPy API to create a CNN image detection and classification model named model_vgg16. The architecture of model_vgg16 is defined in the file VGG16_notop. The VGG16_notop architecture features a final prediction layer with twelve classes.

The model model_vgg16 is configured to process 244 x 244 pixel color images. The computed BGR (Blue, Green, Red) color offset values (as seen in the prior attribute frequency table) for the training data are stored in trainData.channel_means. The image_summary table for the augmented data set shows channel_means values of 157.613 (B), 154.45 (G), and 163.318 (R).

The architecture for model_vgg16 uses pre-trained weights. These weights are stored in a separate file in the project path named VGG_ILSVRC_16_layers.caffemodel.h5.

Because the model_vgg16 value for the include_top setting is False, the imported model weights are loaded for all model layers *except* the final prediction layer.

Note: Advanced and interested readers can obtain the open-source VGG16 model architecture file and pre-trained weight file used in this example from the SAS technical support web. They are contained in the compressed file at https://support.sas.com/documentation/prod-p/vdmml/zip/vgg16.zip.

Example Code 6.14

```
In [14]:
model_vgg16 = VGG16(
                    conn,
                    # VGG16 Model Architecture
                    model_table='VGG16_notop',
                    # 12 Image Classes
                    n_classes=12,
                    # BGR Color Images
                    n_channels=3,
                    # 224 x 224 Image Dimensions
                    width=224,
                    height=224,
                    scale=1,
                    # Read in calculated BGR offsets
                    offsets=trainData.channel_means,
                    # Use Pre-Trained Model Weights
                    pre_trained_weights=True,
                    # Path Spec to Pre-Trained Weight File
                    pre_trained_weights_file=dir_path +
'VGG_ILSVRC_16_layers.caffemodel.h5',
                    include_top=False
                    )
NOTE: Model weights attached successfully!
NOTE: Model table is attached successfully!
NOTE: Model is named to "vgg16_notop" according to the model name in
the table.
```

Now there is a fully defined image detection and classification model using the VGG16 architecture and weights. What does this new model look like? How can you browse it? Use print_summary() to display a table that summarizes the 16 layers in the model_vgg16 model architecture.

Example Code 6.15

```
In [15]:
model_vgg16.print_summary()
```

Layer		Type	Kernel Size	Stride	Activation	Output Size	Number of Parameters
0	data	input	None	None	None	(224, 224, 3)	(0, 0)
1	conv1_1	convo	(3, 3)	1	Rectifier	(224, 224, 64)	(1728, 64)
2	conv1_2	convo	(3, 3)	1	Rectifier	(224, 224, 64)	(36864, 64)
3	pool1	pool	(2, 2)	2	Max	(112, 112, 64)	(0, 0)
4	conv2_1	convo	(3, 3)	1	Rectifier	(112, 112, 128)	(73728, 128)
5	conv2_2	convo	(3, 3)	1	Rectifier	(112, 112, 128)	(147456, 128)
6	pool2	pool	(2, 2)	2	Max	(56, 56, 128)	(0, 0)
7	conv3_1	convo	(3, 3)	1	Rectifier	(56, 56, 256)	(294912, 256)
8	conv3_2	convo	(3, 3)	1	Rectifier	(56, 56, 256)	(589824, 256)
9	conv3_3	convo	(3, 3)	1	Rectifier	(56, 56, 256)	(589824, 256)
10	pool3	pool	(2, 2)	2	Max	(28, 28, 256)	(0, 0)
11	conv4_1	convo	(3, 3)	1	Rectifier	(28, 28, 512)	(1179648, 512)
12	conv4_2	convo	(3, 3)	1	Rectifier	(28, 28, 512)	(2359296, 512)
13	conv4_3	convo	(3, 3)	1	Rectifier	(28, 28, 512)	(2359296, 512)
14	pool4	pool	(2, 2)	2	Max	(14, 14, 512)	(0, 0)
15	conv5_1	convo	(3, 3)	1	Rectifier	(14, 14, 512)	(2359296, 512)
16	conv5_2	convo	(3, 3)	1	Rectifier	(14, 14, 512)	(2359296, 512)
17	conv5_3	convo	(3, 3)	1	Rectifier	(14, 14, 512)	(2359296, 512)
18	pool5	pool	(2, 2)	2	Max	(7, 7, 512)	(0, 0)
19	fc6	fc	(25088, 4096)	None	Rectifier	4096	(102760448, 4096)
20	fc7	fc	(4096, 4096)	None	Rectifier	4096	(16777216, 4096)
21	fc8	output	(4096, 12)	None	Softmax	12	(49152, 12)
22							134309708

Train the Model

Now that the fully defined VGG16 model architecture with initial weights is ready to go, you can train this model with the project's training data. This process begins with the pre-trained weights that were downloaded. The iterative training process enhances and optimizes the model weights for the specific data on which the model is trained.

Use the model fit() command to train the model named model_vgg16. Specify the model parameters. The code below uses the training data table trainData.

The gpu parameter enables the use of GPU2 during processing and offers the capability to pass multiple GPUs. The model uses 32 observations per thread in a mini-batch, using four threads.

The model training uses a learning rate of 0.001, with a maximum of five epochs. Setting log_level=2 configures the SAS output to produce a model iteration history, which can be useful for monitoring fit, loss, and convergence.

Example Code 6.16

```
In [22]:
model_vgg16.fit(# Use trainData table
                data=trainData,
                # Training mini-batch size
                mini_batch_size=32,
                # Only 5 training epochs
                max_epochs=5,
                # Model learning rate
                lr=0.001,
                # Enable GPU processing
                gpu=dict(devices=[2]),
                # Up to 4 threads
                n_threads=4,
                # Level 2 log setting displays
                # training feedback in the output
                log_level=2
                );
```
NOTE: Either dataspecs or inputs need to be non-None, therefore
inputs=_image_ is used
NOTE: Training based on existing weights.
NOTE: Only 1 out of 4 available GPU devices are used.
NOTE: The Synchronous mode is enabled.
NOTE: The total number of parameters is 134309708.
NOTE: The approximate memory cost is 2093.00 MB.
NOTE: Loading weights cost   0.21 (s).
NOTE: Initializing each layer cost   2.76 (s).
NOTE: The total number of threads on each worker is 4.
NOTE: The total mini-batch size per thread on each worker is 32.
NOTE: The maximum mini-batch size across all workers for the
synchronous mode is 128.
NOTE: Target variable: _label_
NOTE: Number of levels for the target variable: 12
NOTE: Levels for the target variable:
NOTE: Level   0: Ant-Man
NOTE: Level   1: Aquaman
NOTE: Level   2: Avengers
NOTE: Level   3: Batman
NOTE: Level   4: Black Panther
NOTE: Level   5: Captain America
NOTE: Level   6: Catwoman
NOTE: Level   7: Ghost Rider
NOTE: Level   8: Hulk
NOTE: Level   9: Iron Man
NOTE: Level   10: Spiderman
NOTE: Level   11: Superman
NOTE: Number of input variables: 1
NOTE: Number of numeric input variables: 1
```

| NOTE: | Epoch | Learning Rate | Loss | Fit Error | Time (s) |
|---|---|---|---|---|---|
| NOTE: | 0 | 0.001 | 1.6894 | 0.5636 | 766.35 |
| NOTE: | 1 | 0.001 | 1.0952 | 0.3583 | 769.52 |
| NOTE: | 2 | 0.001 | 0.7924 | 0.2569 | 770.64 |
| NOTE: | 3 | 0.001 | 0.5897 | 0.1905 | 770.99 |
| NOTE: | 4 | 0.001 | 0.4458 | 0.1436 | 771.81 |

```
NOTE: The optimization reached the maximum number of epochs.
NOTE: The total time is   3849.34 (s).
```
```

Notice that 12 Superheroes are listed in the output classes. There is one class (level) for each Superhero in the target variable.

Now use save() to save the trained model and its enhanced training data in CAS table VGG16_NOTOP_WEIGHTS as super_hero_vgg16_weights. If a model by that name already exists, then the old model is overwritten.

Example Code 6.17

```
In [25]:
# Save Model with Optimized Training Weights
conn.save(table='VGG16_NOTOP_WEIGHTS',
          name='super_hero_vgg16_weights',
          replace=True);
NOTE: Cloud Analytic Services saved the file
super_hero_vgg16_weights.sashdat
in caslib CASUSER(sas).
```

Score the Test Data to Validate Model Accuracy

Now you have a saved, trained image detection and classification model with VGG16 architecture. This trained, weighted model is ready to score new data.

Remember that you partitioned a testData table. That data was not used to train the model, but it contains already-labeled training data that you can score in order to validate model accuracy. Use the predict() command with model_vgg16 to score the model data in the hold back table testData. Two GPUs are enabled for processing. Note that the code to deploy a trained model for scoring is very compact.

Example Code 6.18

```
In [27]:
model_vgg16.predict(# Predict the superhero class for
                    # images in the testData set
                    data=testData,
                    # Use available GPUs
                    gpu=dict(devices=[0,1])
                    )

NOTE: Only 2 out of 4 available GPU devices are used.
```

Out[27]:

§ ScoreInfo

	Descr	Value
0	Number of Observations Read	1086
1	Number of Observations Used	1086
2	Misclassification Error (%)	27.99263
3	Loss Error	0.999919

§ OutputCas Tables

	casLib	Name	Rows	Columns	casTable
0	CASUSER(sas)	Valid_Res_YUsvm3	1086	19	CASTable('Valid_Res_YUsvm3', caslib='CASUSER(s...

elapsed 5.03s · user 10.4s · sys 2.21s · mem 2.51e+03MB

The model scored the data in the test partition and misclassified 304 of 1086 images, resulting in an overall misclassification rate of 28.0%. The remaining 782 images in the test data were correctly classified.

Browse the Scored Data

To see how output scored data appears, you can examine a few example images from the scored data. You can see how the model evaluated each classification. The img_type filter values are C, M, or A. Scored images that are img_type='C' are correctly classified images. Scored images that are img_type='M' are misclassified images. The img_type='A' class returns all scored images, so scored images that are img_type='A' are a combination of both correctly classified and misclassified images. "A" is the "no filter" filter.

The following code uses the results evaluation command plot_evaluate_res() to view the selected images from scored model_vgg16 data, along with a bar chart to show the distribution of the computed target class membership prediction.

Example Code 6.19

```
In [28]:
model_vgg16.plot_evaluate_res(# Show correctly classified images only
                              img_type='C',
                              # Select the images randomly
                              randomize=True,
                              # Select two images
                              n_images=2
                              )

NOTE: Cloud Analytic Services dropped table TEMP_PLOT from caslib
CASUSER(sas).
```

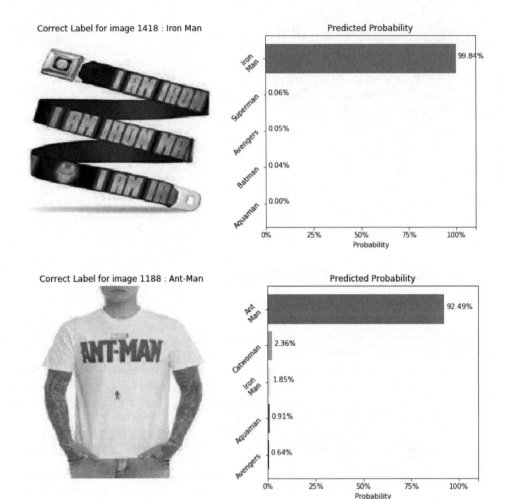

These two images were correctly classified as Iron Man and Ant-Man with a 99.84% and 92.49% probability. Out of fairness, the example should look at some of the images that were misclassified when scored by the model. What do images that are hard to classify look like? Use the evaluate results plot command to show some of the misclassified predictions by the model.

Example Code 6.20

```
In [30]:
model_vgg16.plot_evaluate_res(# Display misclassified images only
                              img_type='M',
                              # Choose images randomly
                              randomize=True,
                              # Display two images
                              n_images=2
                              )
```

```
NOTE: Cloud Analytic Services dropped table TEMP_PLOT from caslib
CASUSER(sas).
```

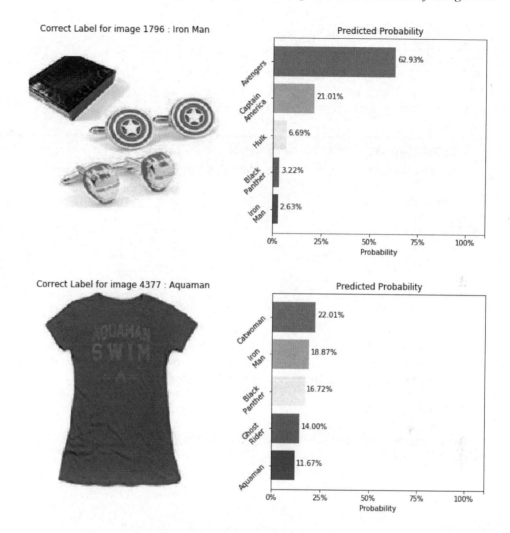

Correct Label for image 1796 : Iron Man

Predicted Probability

Correct Label for image 4377 : Aquaman

Predicted Probability

These two images were misclassified as Avengers (Iron Man) and Cat Woman (AquaMan) with respective probabilities of 62.93% and 22.01%.

Save Model as ASTORE for Deployment

The following code saves the trained and validated VGG16 image detection and classification model named model_vgg16 to the specified path as an ASTORE file.

ASTORE files are binary files that contain the information about the state of an analytic object, such as a predictive deep learning model. ASTORE model files are transportable. The model file can be produced on one host and consumed on another computing environment without the need of traditional SAS export or import utilities.

Example Code 6.21

```
In [36]:
model_vgg16.deploy(path='/data/userID/superhero_classification/',
                   output_format='astore'
                   )

NOTE: Model astore file saved successfully.
```

Upload the ASTORE to SAS CAS

Upload the saved ASTORE file to SAS Cloud Analytic Service (CAS).

Example Code 6.22

```
In [37]:
dlpy.utils.upload_astore(s,

/data/userID/superhero_classification/vgg16_notop.astore',
                         'superhero_vgg16'
                         )
NOTE: Added action set 'astore'.
NOTE: 537256134 bytes were uploaded to the table "superhero_vgg16" "
 in the caslib "CASUSER(sas)".
```

Now you can use the astore.describe CAS action (from the ASTORE CAS action set) to display useful information. You can use this information to specify input-map and output-map parameters to the SAS Event Stream Processing model.

Example Code 6.23

```
In [38]:
s.astore.describe(rstore='superhero_vgg16')
```

```
Out[38]:
```

§ Key

Store Key

	Key
0	8D2EEA1F54A44500B76F04BFF1787AE76829047A

§ Description

Basic Information

	Attribute	Value
0	Analytic Engine	deeplearn
1	Time Created	13Dec2019:11:58:28

§ InputVariables

Input Variables

	Name	Length	Role	Type	RawType	FormatName
0	_image_	29041.0	Input	Varbinary	Character	

§ OutputVariables

Output Variables

	Name	Length	Type	Label
0	P__label_Ant_Man	8.0	Num	Predicted: _label_=Ant-Man
1	P__label_Aquaman	8.0	Num	Predicted: _label_=Aquaman
2	P__label_Avengers	8.0	Num	Predicted: _label_=Avengers
3	P__label_Batman	8.0	Num	Predicted: _label_=Batman
4	P__label_Black_Panther	8.0	Num	Predicted: _label_=Black Panther
5	P__label_Captain_America	8.0	Num	Predicted: _label_=Captain America
6	P__label_Catwoman	8.0	Num	Predicted: _label_=Catwoman
7	P__label_Ghost_Rider	8.0	Num	Predicted: _label_=Ghost Rider
8	P__label_Hulk	8.0	Num	Predicted: _label_=Hulk
9	P__label_Iron_Man	8.0	Num	Predicted: _label_=Iron Man
10	P__label_Spiderman	8.0	Num	Predicted: _label_=Spiderman
11	P__label_Superman	8.0	Num	Predicted: _label_=Superman
12	I__label_	15.0	Character	Into: _label_

elapsed 10.8s · user 10.4s · sys 0.368s · mem 1.31e+03MB

§ Key

Store Key

	Key
0	8D2EEA1F54A44500B76F04BFF1787AE76829047A

§ Description

Basic Information

	Attribute	Value
0	Analytic Engine	deeplearn
1	Time Created	13Dec2019:11:58:28

§ InputVariables

Input Variables

	Name	Length	Role	Type	RawType	FormatName
0	_image_	29041.0	Input	Varbinary	Character	

§ OutputVariables

Output Variables

	Name	Length	Type	Label
0	P__label_Ant_Man	8.0	Num	Predicted: _label_=Ant-Man
1	P__label_Aquaman	8.0	Num	Predicted: _label_=Aquaman
2	P__label_Avengers	8.0	Num	Predicted: _label_=Avengers
3	P__label_Batman	8.0	Num	Predicted: _label_=Batman
4	P__label_Black_Panther	8.0	Num	Predicted: _label_=Black Panther
5	P__label_Captain_America	8.0	Num	Predicted: _label_=Captain America
6	P__label_Catwoman	8.0	Num	Predicted: _label_=Catwoman
7	P__label_Ghost_Rider	8.0	Num	Predicted: _label_=Ghost Rider
8	P__label_Hulk	8.0	Num	Predicted: _label_=Hulk
9	P__label_Iron_Man	8.0	Num	Predicted: _label_=Iron Man
10	P__label_Spiderman	8.0	Num	Predicted: _label_=Spiderman
11	P__label_Superman	8.0	Num	Predicted: _label_=Superman
12	I__label_	15.0	Character	Into: _label_

elapsed 10.8s · user 10.4s · sys 0.368s · mem 1.31e+03MB

Save the testData table as a_.sashdat file for future use and then load the table into CAS.

Example Code 6.24

```
In [39]:
s.table.save(table=testData,
            name='testData',
            caslib='SuperHeroInput',
            replace=True
            )
# Load the table into CAS
s.table.loadtable(path='testData.sashdat',
                caslib='SuperHeroInput',
```

```
                    casOut='TestData'
                    )
```
NOTE: Cloud Analytic Services saved the file testData.sashdat in
caslib SuperHeroInput.
NOTE: Cloud Analytic Services made the file testData.sashdat available
as table TESTDATA in caslib CASUSER(sas).

Out[39]:
§ caslib
CASUSER(sas)

§ tableName
TESTDATA

§ casTable
CASTable('TESTDATA', caslib='CASUSER(sas)')

Print the data that you intend to score. You can scan the output for missing or bad data.

Example Code 6.25

```
In [40]:
# Browse the converted table to look for missing or poorly formatted
# data. This example output has been abbreviated to enhance example
readability.
ScoringInput=s.table.fetch(table='TestData',
                           maxRows=1054,
                           to=1054,
                           sasTypes=False
                           )

ScoringInput
```

```
Out[40]:
```

§ Fetch

Selected Rows from Table TESTDATA

	_filename_0	_image_	_id_	_label_
0	cax_spiderman_train4148.jpg	b'\xff\xd8\xff\xe0\x00\x10JFIF\x00\x01\x01\x00...	4739	Spiderman
1	cax_spiderman_train3874.jpg	b'\xff\xd8\xff\xe0\x00\x10JFIF\x00\x01\x01\x00...	4465	Spiderman
2	cax_spiderman_train4241.jpg	b'\xff\xd8\xff\xe0\x00\x10JFIF\x00\x01\x01\x00...	4832	Spiderman
3	cax_batman_train740.jpg	b'\xff\xd8\xff\xe0\x00\x10JFIF\x00\x01\x01\x00...	1180	Batman
4	cax_ironman_train3436.jpg	b'\xff\xd8\xff\xe0\x00\x10JFIF\x00\x01\x01\x00...	3742	Iron Man
...	...	...	...	...
1049	cax_spiderman_train4310.jpg	b'\xff\xd8\xff\xe0\x00\x10JFIF\x00\x01\x01\x00...	4901	Spiderman
1050	cax_spiderman_train4452.jpg	b'\xff\xd8\xff\xe0\x00\x10JFIF\x00\x01\x01\x00...	5117	Spiderman
1051	cax_spiderman_train4649.jpg	b'\xff\xd8\xff\xe0\x00\x10JFIF\x00\x01\x01\x00...	5507	Spiderman
1052	cax_blackpanther_train1867.jpg	b'\xff\xd8\xff\xe0\x00\x10JFIF\x00\x01\x01\x00...	1869	Black Panther
1053	cax_superman_train4691.jpg	b'\xff\xd8\xff\xe0\x00\x10JFIF\x00\x01\x01\x00...	5565	Superman

1054 rows × 4 columns

elapsed 0.399s · user 0.191s · sys 0.021s · mem 37.1MB

SAS Event Stream Processing analytics requires images that are formatted as base64-encoded strings. Here we convert the image data to be scored into a base64-encoded string.

Example Code 6.26

```python
In [41]:
# Import Python analytic mathematical libraries
# Python Base64 encoding library
import base64
# Python Computer Vision library
import cv2
# Python scientific computing library
import numpy as np
# Python data structures and analysis library
import pandas as pd
# Custom Python function to convert binary fields to Base64-
# encoded strings for SAS ESP analytics compatibility
def bst2b64(df, binaryfields=[], keepfields = []):
    """ Custom function that converts binary fields to Base64. Formats
the input to be compatible with SAS Event Stream Processing.
    """
    df2 = df.copy()
    for col in df2.columns:
        if col in binaryfields:
            newname = "b64_"+col
            tmp_col = []
            for row in df2[col]:
                tmp_col.append(base64.b64encode(row).decode('utf-8'))
            df2.drop(col,
                    axis = 1,
                    inplace = True
```

```
                    )
            df2[newname] = tmp_col
            keepfields.append(newname)
            if col in keepfields:
                try:
                    keepfields.remove(col)
                except:
                    print("{} didn't exist".format(col))
    return df2[keepfields]
In [42]:
# Import the SAS CAS TestData table into Pandas dataframe
ImagesDF = pd.DataFrame(s.fetch(table=s.CASTable('TestData'),
                            maxRows=1054,
                            to=1054
                            )
                    ['Fetch']
                    )
# Convert UTF-8 binaries into 64-bit images
# Preserve original image ID and image label
b64_out = bst2b64(ImagesDF,
                binaryfields=['_image_'],
                keepfields=['_image_',
                            '_id_',
                            '_label_']
                )

# Rename the columns in the converted table
# using appropriate column headings
ImageData_ESP_Input=b64_out.rename(columns={"_id_":"_id_",

                                    "_label_":"_label_",

                                        "b64__image_":"_image_"
                                        }
                                )
# Display the newly converted 64-bit encoded table,
# Browse the table to look for missing or poorly formatted
# data. This example output shown below is condensed in order
# to enhance the readability of this example.
ImageData_ESP_Input
```

```
Out[42]:
```

	_id_	_label_	_image_
0	4739	Spiderman	/9j/4AAQSkZJRgABAQAAAQABAAD/2wBDAAIBAQEBAQIBAQ...
1	4465	Spiderman	/9j/4AAQSkZJRgABAQAAAQABAAD/2wBDAAIBAQEBAQIBAQ...
2	4832	Spiderman	/9j/4AAQSkZJRgABAQAAAQABAAD/2wBDAAIBAQEBAQIBAQ...
3	1180	Batman	/9j/4AAQSkZJRgABAQAAAQABAAD/2wBDAAIBAQEBAQIBAQ...
4	3742	Iron Man	/9j/4AAQSkZJRgABAQAAAQABAAD/2wBDAAIBAQEBAQIBAQ...
...	...	...	...
1049	4901	Spiderman	/9j/4AAQSkZJRgABAQAAAQABAAD/2wBDAAIBAQEBAQIBAQ...
1050	5117	Spiderman	/9j/4AAQSkZJRgABAQAAAQABAAD/2wBDAAIBAQEBAQIBAQ...
1051	5507	Spiderman	/9j/4AAQSkZJRgABAQAAAQABAAD/2wBDAAIBAQEBAQIBAQ...
1052	1869	Black Panther	/9j/4AAQSkZJRgABAQAAAQABAAD/2wBDAAIBAQEBAQIBAQ...
1053	5565	Superman	/9j/4AAQSkZJRgABAQAAAQABAAD/2wBDAAIBAQEBAQIBAQ...

1054 rows × 3 columns

Use SAS ESPPy to Deploy ASTORE Model and Score Streaming Data

SAS ESPPy is an open-source package for Python that is available
at https://github.com/sassoftware/python-esppy. With SAS ESPPy, you can create SAS Event
Stream Processing models programmatically, connect to an ESP server, and interact with
projects and their components as Python objects. These objects include continuous queries,
windows, events, loggers, SAS Micro Analytic Service modules, routers, and analytical
algorithms. SAS ESPPy fully integrates with the JupyterLabs toolset, which supports
visualizing diagrams of your ESP projects and streaming charts.

Example Code 6.27

```
In [43]:
# Import esppy Python module and ESP library
Import esppy
# Use ESPPY to connect to an existing ESP server
esp = esppy.ESP('http://<host>:portnum')

In [44]:
# Create a SAS ESP project named 'project_01' and
# create a continuous query named 'cq_01'
proj = esp.create_project('project_01',
                          pubsub='auto',
                          n_threads='1'
                          )
cq = esp.ContinuousQuery(name='cq_01',
                         trace='w_score'
                         )
```

```
In [45]:
# Create an ESP Source Window and define
# a connector to publish the image data
src = esp.SourceWindow(name='w_data',
                       insert_only=True,
                       autogen_key=True,
                       schema=('index*:int64',
                               '_id_:double',
                               '_label_:string',
                               '_image_:blob'
                               )
                       )
# Display the source window
src

Out[45]:
```

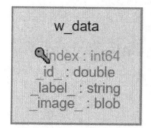

Next, create a Model Reader Window and define parameters to pull in the ASTORE file.

Example Code 6.28

```
In [46]:
# Create Model Reader Window for ASTORE
w_reader = esp.ModelReaderWindow(name='w_reader',
                                 model_type='astore'
                                 )
# Specify Model Reader Window Parameters
w_reader.set_parameters(action='load',
                        type='astore',
                        reference=
'/data/userID/superhero_classification/vgg16_notop.astore'
                        )
# Display the ASTORE Model Reader Window
w_reader

Out[46]:
```

w_reader

Now, create a Score window. The Score window is where the scored ESP data will be published. This is where you specify the mapping for the input_map and output_map. Use the mapping information that was contained in the output from the *astore.describe* CAS action that the example ran earlier.

Example Code 6.29

```
In [47]:
# Define the initial display schema for the Score window:
# Display predictive score values
w_score = esp.ScoreWindow(name='w_score',
# Display output schema for 12 superhero target classes
                          schema=('index*:int64',
                                  'P__label_Ant_Man:double',
                                  'P__label_Aquaman:double',
                                  'P__label_Avengers:double',
                                  'P__label_Batman:double',
                                  'P__label_Black_Panther:double',
                                  'P__label_Captain_America:double',
                                  'P__label_Catwoman:double',
                                  'P__label_Ghost_Rider:double',
                                  'P__label_Hulk:double',
                                  'P__label_Iron_Man:double',
                                  'P__label_Spiderman:double',
                                  'P__label_Superman:double',
                                  'I__label_:string'
                                  )
                          )
# Attach the ASTORE model to the score window by
# specifying the appropriate input-map and output-map:
w_score.add_offline_model(model_type='astore',
# Specify Input Map Parameters
                          input_map={'_image_':'_image_'},
# Specify Output Map Parameters
                          output_map={
                          'P__label_Ant_Man':'P__label_Ant_Man',
                          'P__label_Aquaman':'P__label_Aquaman',
                          'P__label_Avengers':'P__label_Avengers',
                          'P__label_Batman':'P__label_Batman',

'P__label_Black_Panther':'P__label_Black_Panther',

'P__label_Captain_America':'P__label_Captain_America',
                          'P__label_Catwoman':'P__label_Catwoman',

'P__label_Ghost_Rider':'P__label_Ghost_Rider',
                          'P__label_Hulk':'P__label_Hulk',
                          'P__label_Iron_Man':'P__label_Iron_Man',
                          'P__label_Spiderman':'P__label_Spiderman',
                          'P__label_Superman':'P__label_Superman',
                          'I__label':'I__label'})
# Show a display of the completed ASTORE model
# score window.
w_score
```

```
Out[47]:
```

```
                    w_score

              index : int64
       P__label_Ant_Man : double
       P__label_Aquaman : double
       P__label_Avengers : double
        P__label_Batman : double
     P__label_Black_Panther : double
   P__label_Captain_America : double
      P__label_Catwoman : double
     P__label_Ghost_Rider : double
         P__label_Hulk : double
       P__label_Iron_Man : double
      P__label_Spiderman : double
      P__label_Superman : double
          I__label_ : string
```

Build the SAS Event Stream Processing project by connecting the project definition and continuous query. Then append each of the windows to the continuous query.

Example Code 6.30

```
In [48]:
proj.add_query(cq)
cq.add_window(src)
cq.add_window(w_reader)
cq.add_window(w_score)
# Finish the creation of the ESP project by
# defining edges to connect the windows:
src.add_target(w_score,
               role='data'
               )
w_reader.add_target(w_score,
                    role='model'
                    )
# Print a completed visualization of the project
# named project_01.
 proj
```

Out[48]:

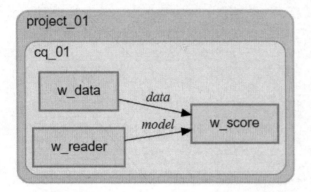

Now we can load the project into the ESP server and begin publishing data into the project.

Example Code 6.31
```
In [49]:
esp.load_project(proj)
# Subscribe to both the source window and the score window
# to verify data is correctly passing through our model.
src.subscribe()
w_score.subscribe()
# Start the SAS ESP project and publish
# the testing images to the source window.
proj.start()
src.publish_events(ImageData_ESP_Input,
                    pause=200
                    )
```

As the model in the SAS ESP project begins to run, you can watch the source window populate with input events. Each time you print the contents of the source window, more and more events appear. The output below has been abbreviated to enhance reading of this example code.

Example Code 6.32
```
In [50]:
src
```

```
Out[50]:
```

index	_id_	_label_	_image_
0	4739.0	Spiderman	b'\xff\xd8\xff\xe0\x00\x10JFIF\x00\x01\x01\x00...
1	4465.0	Spiderman	b'\xff\xd8\xff\xe0\x00\x10JFIF\x00\x01\x01\x00...
2	4832.0	Spiderman	b'\xff\xd8\xff\xe0\x00\x10JFIF\x00\x01\x01\x00...
3	1180.0	Batman	b'\xff\xd8\xff\xe0\x00\x10JFIF\x00\x01\x01\x00...
4	3742.0	Iron Man	b'\xff\xd8\xff\xe0\x00\x10JFIF\x00\x01\x01\x00...
...	...	...	...
1049	4901.0	Spiderman	b'\xff\xd8\xff\xe0\x00\x10JFIF\x00\x01\x01\x00...
1050	5117.0	Spiderman	b'\xff\xd8\xff\xe0\x00\x10JFIF\x00\x01\x01\x00...
1051	5507.0	Spiderman	b'\xff\xd8\xff\xe0\x00\x10JFIF\x00\x01\x01\x00...
1052	1869.0	Black Panther	b'\xff\xd8\xff\xe0\x00\x10JFIF\x00\x01\x01\x00...
1053	5565.0	Superman	b'\xff\xd8\xff\xe0\x00\x10JFIF\x00\x01\x01\x00...

1054 rows × 3 columns

After you use the deployed model to publish and score the input images, you can use the **w.score** command to create output that shows the class distribution of the scored image targets. The score window continues to populate incrementally as more image scoring events are processed. The score command is displayed below. For the convenience of the reader, the resulting output chart with 1,053 rows and 13 columns is omitted.

Example Code 6.33

```
In [52]:
w_score
```

You can save the project in an XML file and create a portable model. This enables you to load the portable model directly from the XML file in a future environment. The code to do so resembles the following:

Example Code 6.34

```
In [45]:
proj.save_xml('SuperHeroModel.xml',
              mode='w'
             )
```

After you finish your work, it is a good practice to stop the ESP server.

Example Code 6.35

```
In [53]:
esp.shutdown()
```

What Will the Future Bring?

According to IDC and Forrester, by 2024, AI will be integral to *every* part of a business. AI will become the new UI, redefining user experiences with technology. Over 50% of user touches will be amplified by computer vision, speech, natural language, augmented reality, and virtual reality. On the immediate horizon, AI and the emerging user interfaces of computer vision, natural language processing, and gesture are going to be embedded in every type of product and device (Press 2019).

AI applied within the IoT will change how we work in health care, manufacturing, and product delivery. AI will be embedded in "ordinary" home systems as well: automation, security, entertainment, and energy management. As new tools and technology are adapted, we must prepare for the societal changes as well as the ethical challenges that we will encounter while managing the tremendous amount of metadata generated by increasing numbers of embedded IoT devices.

IoT Governance

The expanding growth of IoT devices will generate a proportionally large cohort of user and trust domain metadata. As the technology becomes more broadly distributed and enjoyed, there are downside risks to consider. Consumers who deploy smart technology without engaging appropriate security measures can become vulnerable to damaging personal data theft or misuse.

Who owns the data collected by your smart sensor systems? How will it be secured for privacy? What happens when collections of personal, financial, medical, or security metadata are transmitted or stored on wireless networks with inferior security protection? Are there minimum standards for IoT device and data security that can prevent misappropriation, theft, or abuse of private data?

The international community is already responding to the cultural and informational changes brought about by IoT. On May 25, 2018, The European Union (EU) began enforcing a General Data Protection Regulation (GDPR) to ensure the security and privacy of citizens in the EU and the European Economic Area (EEA). The intent of GDPR is to provide individuals control over their personal data and to simplify international business requirements by providing the EU with a common set of provisions and requirements for processing, storage, ownership, and conditional use of personal data. For more information about the EU GDPR, see https://eugdpr.org.

The GDPR regulation also addresses the transfer of data in locations that are outside of the EU and EEA areas. The United States does not have an equivalent to the EU GDPR regulations, but US companies that control or process the personal data of persons living in EU or EEA zones are required to comply with all GDPR provisions and regulations. US company CIOs and CTOs with EU customers must plan for the scope of GDPR compliance and digital infrastructure requirements.

5G Networking

IoT architects and device manufacturers design networks in order to balance sets of competing requirements: power consumption and power source, operating cost, manufacturing cost, network range, network latency and bandwidth, network signal quality, and network node density, to name a few. Currently there is no single networking technology that can meet and optimize these competing requirements. Combinations of different network deployments abound, giving information architects more flexibility and choice in system design. But there are many factors to recommend 5G networks and backscatter networks.

Experts are calling 5G mobile networks the next-generation wireless communication standard. 5G networks have begun to take over mobile networks in 2019, but 5G will not immediately replace LTE networks and their variants. 5G networks are attractive to IoT devices because they use a high-frequency band of the wireless spectrum between 28 and 60GHz, known as the millimeter wave spectrum. In addition to increased bandwidth available to users, 5G networks deliver lower latency, increased capacity, and faster transmission speeds.

Backscatter networks are a low-power communication technology that use existing radio frequency (RF) signals (such as radio, TV, and mobile phones) as a power source. (See Figure 6.5.) They reuse existing Wi-Fi infrastructures to transmit data and provide RF-powered devices with internet connectivity. The technology uses less power than Wi-Fi because it piggybacks signals on existing Wi-Fi devices. Backscattering refers to the practice of selectively reflecting existing RF signals to recharge battery-powered sensors, or even eliminating the need for a sensor battery in some cases. The reflected RF signals create a pattern of stronger and weaker signals that can be detected by specially tuned Wi-Fi routers.

Figure 6.5: Ultra-Wideband Ambient Backscatter System

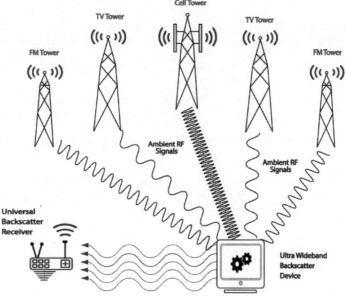

Backscatter networks are a good fit for IoT applications because the low-bit rate network capacities and low power requirements work well with IoT sensor arrays, machine-to-machine networks, as well as RF computing.

Conclusion

So how would Ray Bradbury react to our present-day Internet of Things? What type of science fiction would he invent to extrapolate on the advances that we have made in the seventy years since the publication of "There Will Come Soft Rains?" On the one hand, it is easy to envision intelligent devices getting smaller, more embeddable, and more pervasive, continuing the trend of the past ten years. On the other hand, it is difficult to imagine what completely new applications for AI at the edge will be invented. These days, it is easy to imagine the extension of smart phone tracking within 5G networks, but 20 years ago, who could have imagined smart phones? These days, it is easy to imagine the extension of strong AI into new areas, but who can imagine what type of artificial intelligence AI will invent?

Perhaps we should leave the answers to those questions to writers such as Bradbury. Still, there are three things that we can be sure of, at least for the near future: the data center will become embedded in the cloud, AI will become more adaptive and pervasive, and everywhere that we go will be at the edge.

References

Bradbury, R. "There Will Come Soft Rains" (May 6, 1950), Collier's Weekly, New York, NY.

Chauhan, A. S. "How AI is transforming Home Automation" (September 18, 2018), Becoming Human: Artificial Intelligence Magazine, Medium.com. https://becominghuman.ai/how-ai-is-transforming-home-automation-56085cb275b

Gartner, Inc. Newsroom Press Release, "Top 10 Strategic IoT Technologies and Trends" (November 7, 2018), Gartner Symposium, Barcelona, Spain. https://www.gartner.com/en/newsroom/press-releases/2018-11-07-gartner-identifies-top-10-strategic-iot-technologies-and-trends

Kumar, C. "Artificial Intelligence: Definition, Types, Examples, Technologies" (August 2018), Medium.com. https://medium.com/@chethankumargn/artificial-intelligence-definition-types-examples-technologies-962ea75c7b9b

Laney, D. B. Infonomics: How to Monetize, Manage, and Measure Information as an Asset for Competitive Advantage. New York: Gartner, 2018.

Marr, B. "What is Digital Twin Technology – And Why Is It So Important?" (March 6, 2017), Forbes Magazine. https://www.forbes.com/sites/bernardmarr/2017/03/06/what-is-digital-twin-technology-and-why-is-it-so-important/#5b6b442c2e2a

Pope, D. "Want real-time predictive analysis? You need Event Stream Processing." SAS Voices (blog). January 22, 2014. https://blogs.sas.com/content/sascom/2014/01/22/want-real-time-predictive-analysis-you-need-event-stream-processing/

Press, G. "Top Artificial Intelligence (AI) Predictions For 2020 From IDC and Forrester" (November 22, 2019), Forbes Magazine. https://www.forbes.com/sites/gilpress/2019/11/22/top-artificial-intelligence-ai-predictions-for-2020-from-idc-and-forrester/#4a9d0229315a

About the Contributor

Chip Robie is a SAS Principal Technical Writer with 28 years of experience in data mining, advanced analytics, and deep learning. Prior to SAS, he designed SCADA networks and predictive maintenance systems for an international Fortune 500 company before becoming practice general manager for an international management operations consulting company. He has been published in national technical and business journals, as well as a series of instructional SAS analytics videos.

Acknowledgment

The authors would like to acknowledge Amanda Goldman for her assistance in developing and testing the example included in this chapter.

Chapter 7: Using Geofences with SAS Event Stream Processing

By Michael Harvey and Frédéric Combaneyre

What Is a Geofence?

A *geofence* is a virtual perimeter for a real-world geographic area. You can dynamically generate a geofence as a radius around a specific location or create one as a set of specific boundaries. When a person, vehicle, or device crosses the boundary that you set, the geofence issues an alert. The alert can be configured to report the location of the thing that crossed the boundary, prompt mobile push notifications, trigger text messages, send targeted advertisements on social media, or deliver location-based marketing data.

Thus, with geofences, you can enter a shopping mall and immediately receive commercial ads and offers based on your personal taste and past purchases. Authorities can track vessels' positions, detect when a ship is not in the area where it should be, or forecast and optimize harbor arrivals. When a truck driver breaks from a route, the dispatcher can be alerted and can act immediately. There are countless other examples from manufacturing, industry, security, or even households.

Common geofencing applications include the following (White 2017):

- **Social networking**. Location-based filters and other shareable content are made possible through geofencing.
- **Marketing**. Businesses can use geofencing to deliver in-store promotions, alerting customers as they step in range of the store.
- **Audience engagement**. Geofencing can be used to engage crowds of people at organized events like concerts, festivals, and fairs to deliver information about the venue or event.
- **Smart appliances**. When an appliance is connected to the internet, a geofence can provide a timely alert to a change of state (for example, your refrigerator can remind you that you are out of milk as you pass by the grocery store).
- **Human resources**. Companies can use geofencing to monitor workers who work off-site. They can also automate time cards, clocking employees in and out as they arrive at and leave the office.
- **Telematics**. Geofencing can draw virtual zones around sites, work areas, and secure areas.
- **Security**. Geofences can be used to unlock your phone when you are home or get alerts when someone enters or leaves your house.

Understanding the Geofence Window

The Geofence window that is provided by SAS Event Stream Processing determines whether the location of an event stream is inside or close to an area of interest. What follows is a general overview of the behavior of the Geofence window. For a complete explanation, please refer to the SAS Event Stream Processing documentation.

Geofence windows require two input windows: one to inject streaming events (the position source window) and another to inject geofence areas and locations (the geometry source window). By default, you connect the window that injects events to the Geofence window with the first edge that you specify in the project. You connect the window that injects geofence areas and locations to the Geofence window with the second edge.

Example Code 7.1

```
<edges>
  <edge source='position_source' target='geofence_window'/>
  <edge source='geometry_source' target='geofence_window'/>
</edges>
```

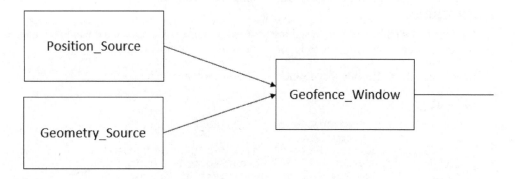

Thus, a Geofence window behaves like an outer Join window or a lookup operation. Events are on the streaming side and the geofence areas and locations are on the dimension side.

Alternatively, you can explicitly assign a role to the edges that connect input windows to Geofence Windows: position or geometry. For example:

Example Code 7.2

```
<edges>
  <edge source='geometry_source' target='geofence_window'
    role='geometry'/>
  <edge source='position_source' target='geofence_window'
    role='position'/>
</edges>
```

The Geofence window is designed to support Cartesian or geographic coordinate types. The only requirement is that all coordinates are consistent and refer to the same space or projection. For geographic coordinates, the coordinates must be specified in the (X,Y) Cartesian order (longitude, latitude). All distances are defined and calculated in meters.

The output schema of a Geofence window is automatically defined by the following fields in the following order:

- All fields that come from the position source window in their respective order.
- A mandatory field of type int64 or string that receives the ID of the geometry. If no geometries are found, then the value of this field is null in the produced event. This field is defined by the parameter geoid-fieldname.

Geometries

Geometries are areas and locations of interest. The Geofence window supports the following geometries: polygons, circles, and polylines. One event is published per geometry.

The Geofence window supports Insert, Update, and Delete opcodes, which can dynamically update the geometries.

Polygons

A *polygon* is a plane shape representing an area of interest. The Geofence window supports polygons, multi-polygons, and polygons with holes or multiple rings.

Here is sample XML code that defines a polygon geofence:

Example Code 7.3

```
<window-geofence name="geofence_poly" index="pi_EMPTY"
                  output-insert-only="true">
    <geofence coordinate-type="geographic" meshfactor-x="-2"
              meshfactor-y="-2" output-sorted-results="true"
              max-meshcells-per-geometry="10" autosize-mesh="true"/>
    <geometry data-fieldname="poly_data" desc-fieldname="poly_desc"/>
    <position x-fieldname="GPS_longitude" y-fieldname="GPS_latitude"/>
    <output geoid-fieldname="poly_id" geodesc-fieldname="poly_desc"
            geodistance-fieldname="poly_dist"/>
</window-geofence>
```

The <geofence> element specifies the general properties of the Geofence window. In this window, geographic coordinates are used. The specified mesh factors for the X or longitude axis and the Y or latitude axis are fed to a mesh algorithm. This algorithm uses a parameter (called a *mesh factor*) that defines the scale of the space subdivision.

The mesh factor is an integer within the range [-5, 5] that represents a power of 10 of the coordinate units in use. The default value is 0. That default value generates 1 subdivision per coordinate unit. A value of 1 generates a subdivision per 10 units. A value of -1 generates 10 subdivisions per unit. The mesh factor can be set for both X and Y axes or independently for each axis. The mesh factor is explained in detail in the SAS Event Stream Processing documentation.

Because the output-sorted-results attribute of the <geofence> element is set to true, the output result is sorted by increasing distance between the position and the geometry. The max-meshcells-per-geometry parameter of that element specifies the maximum allowed mesh cells created per geometries. This avoids creating an oversized mesh that would generate unnecessary calculations. Whenever a geometry exceeds the specified limit, it is rejected.

The <position> element specifies window positioning. This window specifies the name of the position input window's fields that contain the position X or longitude coordinate and the position Y or latitude coordinate.

The <output> element specifies the structure of Geofence window output. Here, the geoid-fieldname attribute specifies the name of the output schema field that receives the geometry ID. The geodesc-fieldname attribute specifies the name of the output schema field that receives the geometry description. The geodistance-fieldname attribute specifies the name of the output schema field that receives the distance between the position and the geometry, in this case, the centroid for the polygons.

Circles

A *circle* encompasses the position of a location of interest. Define a circle with three values:

- two coordinates, X and Y (longitude and latitude), that represent the center of the circle
- a radius distance around the center

Here is sample XML code that specifies a circle:

Example Code 7.4

```
<window-geofence name="geofence_circles" index="pi_EMPTY"
                output-insert-only="true">
    <geofence coordinate-type="geographic" output-sorted
     results="true" meshfactor-x="-2" meshfactor-y="2"
     max-meshcells-per-geometry="10" autosize-mesh="true"/>
    <geometry x-fieldname="POI_x" y-fieldname="POI_y" desc
     fieldname="POI_desc" radius-fieldname="POI_radius"/>
    <position x-fieldname="GPS_longitude" y-fieldname="GPS_latitude"
     lookupdistance="110"/>
    <output geoid-fieldname="POI_id" geodesc-fieldname="POI_desc"
     eventnumber-fieldname="event_nb" geodistance
     fieldname="POI_dist"/>
</window-geofence>
```

Note the similarities to the XML code that specified a polygon. As with the polygon, the geodistance-fieldname attribute specifies the name of the output schema field that receives the distance between the position and the geometry. However, in this case, the center point for the circle is specified.

Polylines

A *polyline* is a customized shape that represents a border or a trip wire. You define a polyline as a list of coordinates that represent the polyline's segment or segments. The sequence of the points in the segment definition defines the polyline's vector direction and its left and right side.

When working with polylines, a Geofence window analyzes each event position that comes from the position source window. It returns the ID of the polyline that is within the distance defined by the radius property value. Distance is measured from the position to the closest segment.

Consider Figure 7.1. The polyline is depicted as a bold black line. The dashed line around the polyline is the area within the distance of the radius. A, B, C, and D are incoming event positions.

Figure 7.1: Polyline Relative to Several Event Positions

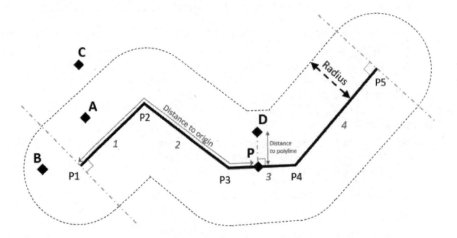

The distance between event position A and the polyline is less than the radius. For that event, the Geofence window returns the polyline ID.

The event position B is close to the polyline when the strict-projection attribute of the geofence element 'false'. In that case, the polyline ID is returned. B is out of the polyline proximity when strict-projection='true'. The polyline ID is not returned.

The event position C is out of the polyline proximity. The polyline ID is not returned.

With regard to event position D:

- P is the projected point of position D. The coordinates of P are returned by the fields specified by the projection-fieldname parameter of the output element.
- The distance between D and P is returned by the field specified by the geodistance-fieldname parameter of the output element.
- P to P3 to P2 to P1 is the distance to origin. That distance is returned by the field specified by the distance-to-origin-fieldname parameter of the output element.
- The segment number is 3, which is returned by the field specified by the segmentnumber-fieldname parameter of the output element.

Example

The following Jupyter notebook contains a Python program that demonstrates how to use SAS Event Stream Processing to use geofencing on a streaming project. It tracks a vehicle through the streets of Paris, France.

This notebook is available to everyone through the free trial ESP programming environment in the SAS Analytics Cloud.

Set Up the Environment

First, import the necessary packages to run the notebook.

- ipywidgets provides a set of eventful Python objects that have a representation in a browser (for example, a slider or text box).
- ESPPy enables you to create SAS Event Stream Processing models programmatically in Python.
- The Visuals package, which is part of ESPPy, enables you to create various charts, an event stream processing model viewer, and a log viewer with a set of programming objects.
- csv is a module that implements classes to read and write tabular data in CSV format.
- requests is an Apache2 licensed library that enables you to make HTTP requests in Python. You can use it to add query strings to URLs or form-encode POST data.
- The time module provides various time-related functions.
- The threading module enables you to run multiple threads (tasks, function calls) at the same time in your Python program.
- The datetime module enables you to work with dates as date objects.
- The io module provides facilities to deal with various types of I/O in your Python program.

Additionally, set the home directory, which tells the notebook where to save the model XML file.

Example Code 7.5

```
In [ ]:
import ipywidgets as widgets
import esppy
from esppy.espapi.visuals import Visuals
import csv
import requests
import time
import threading
import datetime
import io

from os import path
homedir = path.expanduser("~")
```

Next, create an ESP server object (using esppy.ESP) and create a ServerConnection object (using esp.createServerConnection) to create data sources that feed visualizations.

Example Code 7.6

```
In [ ]:
esp = esppy.ESP('http://localhost:5001')
conn = esp.createServerConnection(interval=0)
esp
```

Create a Class to Contain GPS Data

Next, create a Python class named *MyPublisher* that processes GPS data (such as speed, longitude, and latitude) of the vehicle. The instantiation of this class is, in effect, the vehicle that traverses the streets of Paris. It uses a Publisher object to publish events into a Source window (in this case, the window named position_in, which is described later in the chapter). The GPS data is contained in a CSV file named gf_paris_trip.csv.

Example Code 7.7

```
In [ ]:
class MyPublisher(object):
    def __init__(self,vehicle):
        self._vehicle = vehicle
        threading.Thread(target = self.publish,daemon=True).start()

    def publish(self):
        publisher =
conn.getPublisher("p/cq1/position_in",opcode="insert")
        fields = ["opcode","flag","vehicle","pt_id","GPS_longitude",
                "GPS_latitude","speed","course"]
        with
open("/demo/Event_Stream_Processing/data/gf_paris_trip.csv") as
csvfile:
            csv_reader = csv.reader(csvfile)

            for row in csv_reader:
                publisher.begin()
                for i,x in enumerate(row):
                    if fields[i] == "vehicle":
                        publisher.set(fields[i],self._vehicle)
                    else:
                        publisher.set(fields[i],str(x))
                publisher.set("time",datetime.datetime.now())
                publisher.end()
                publisher.publish()
                #time.sleep(.4)
                time.sleep(.3)

        publisher.close()
```

Now use the Visuals package to specify the color and look of the map of Paris.

Example Code 7.8

```
In [ ]:
visuals = Visuals(colormap="sas_marine",border="1px solid #d8d8d8")
```

Load the SAS Event Stream Processing Project

Load a project named p that contains a continuous query named cq to the ESP server.

Example Code 7.9

```
In [ ]:
esp.load_project("/demo/Event_Stream_Processing/data/gf_model.xml",nam
e="p",overwrite=True)
Out []:
```

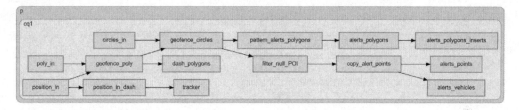

The position source window is named position_in. It is the Source window into which the Publisher object publishes events about the vehicle.

Example Code 7.10

```
<window-source name="position_in" pubsub="true" insert-only="true"
collapse-updates="true" autogen-key="true">
        <schema>
           <fields>
              <field name="vehicle" type="string"/>
              <field name="pt_id" type="int64" key="true"/>
              <field name="GPS_longitude" type="double"/>
              <field name="GPS_latitude" type="double"/>
              <field name="speed" type="double"/>
              <field name="course" type="double"/>
              <field name="time" type="stamp"/>
           </fields>
        </schema>
</window-source>
```

The position source window feeds two windows. The first is a Copy window named position_in_dash. The retention policy is bytime_sliding with deletes automatically generated every 1 second.

Example Code 7.11

```
<window-copy name="position_in_dash" collapse-updates="true"
index="pi_HASH">
        <retention type="bytime_sliding">1 seconds</retention>
</window-copy>
```

The other window fed by the position source window is a Geofence window named geofence_poly, which is described later.

One geometry source window is named poly_in. It uses a file and socket connector to read the file gf_poly_in.csv.

Example Code 7.12

```
<window-source name="poly_in" pubsub="true" index="pi_HASH" insert-
only="true" collapse-updates="true">
        <schema>
          <fields>
            <field name="poly_id" type="int64" key="true"/>
            <field name="poly_desc" type="string"/>
            <field name="poly_data" type="string"/>
          </fields>
        </schema>
        <connectors>
          <connector name="polygons_in" class="fs">
            <properties>
              <property name="type">pub</property>
              <property name="fsname">
               /demo/Event_Stream_Processing/data/gf_poly_in.csv
              </property>
              <property name="fstype">csv</property>
            </properties>
          </connector>
        </connectors>
</window-source>
```

The Geofence window that processes the polygon geometry is named geofence_poly.

Example Code 7.13

```
<window-geofence name="geofence_poly" index="pi_EMPTY"
                 output-insert-only="true">
  <geofence coordinate-type="geographic" meshfactor-x="-2"
    meshfactor-y="-2" log-invalid-geometry="false"
    output-multiple-results="false" output-sorted-results="true"
    max-meshcells-per-geometry="10" autosize-mesh="true"/>
  <geometry data-fieldname="poly_data" desc-fieldname="poly_desc"/>
  <position x-fieldname="GPS_longitude" y-fieldname="GPS_latitude"/>
  <output geoid-fieldname="poly_id" geodesc-fieldname="poly_desc"
          geodistance-fieldname="poly_dist"/>
</window-geofence>
```

Another geometry source window is named circles_in.

Example Code 7.14

```
<window-source name="circles_in" index="pi_HASH" pubsub="true" insert-
only="true" collapse-updates="true">
        <schema>
          <fields>
            <field name="POI_id" type="int64" key="true"/>
            <field name="POI_x" type="double"/>
            <field name="POI_y" type="double"/>
            <field name="POI_desc" type="string"/>
            <field name="POI_radius" type="double"/>
          </fields>
        </schema>
        <connectors>
          <connector name="circles_in" class="fs">
            <properties>
              <property name="type">pub</property>
```

```
              <property name="fsname">
              /demo/Event_Stream_Processing/data/gf_circles_in.csv
              </property>
              <property name="fstype">csv</property>
            </properties>
          </connector>
        </connectors>
      </window-source>
```

The Geofence window that processes the circle geometry is named geofence_circles.

Example Code 7.15

```
<window-geofence name="geofence_circles" index="pi_EMPTY"
                 output-insert-only="true">
  <geofence coordinate-type="geographic" log-invalid-geometry="false"
   output-multiple-results="false" output-sorted-results="true"
   meshfactor-x="-2" meshfactor-y="-2" max-meshcells-per-geometry="10"
   autosize-mesh="true"/>
  <geometry x-fieldname="POI_x" y-fieldname="POI_y" desc-
    fieldname="POI_desc" radius-fieldname="POI_radius"/>
  <position x-fieldname="GPS_longitude" y-fieldname="GPS_latitude"
    lookupdistance="110"/>
  <output geoid-fieldname="POI_id" geodesc-fieldname="POI_desc"
    eventnumber-fieldname="event_nb" geodistance-fieldname="POI_dist"/>
</window-geofence>
```

The window geofence_poly feeds geofence_circles and a Copy window named dash_polygons, which has the same retention policy as the previous Copy window.

Example Code 7.16

```
<window-copy name="dash_polygons" collapse-updates="true"
index="pi_HASH">
            <retention type="bytime_sliding">1 seconds</retention>
</window-copy>
```

The window position_in_dash feeds an Aggregate window named tracker, which performs a number of calculations.

Example Code 7.17

```
<window-aggregate name="tracker" index="pi_HASH">
        <schema>
          <fields>
            <field name="vehicle" type="string" key="true"/>
            <field name="speed" type="double"/>
            <field name="course" type="double"/>
            <field name="GPS_longitude" type="double"/>
            <field name="GPS_latitude" type="double"/>
          </fields>
        </schema>
        <output>
          <field-expr>
            <![CDATA[ ESP_aLast(speed) ]]>
          </field-expr>
          <field-expr>
            <![CDATA[ ESP_aLast(course) ]]>
```

```
            </field-expr>
            <field-expr>
              <![CDATA[ ESP_aLast(GPS_longitude) ]]>
            </field-expr>
            <field-expr>
              <![CDATA[ ESP_aLast(GPS_latitude) ]]>
            </field-expr>
          </output>
  </window-aggregate>
```

The window geofence_circles feeds a Pattern window named pattern_alerts_polygons. This window processes the geofence data as input and builds events of interest (EOIs) that correspond to positions that cross the geometry boundary.

Example Code 7.18

```
<window-pattern name="pattern_alerts_polygons">
        <schema>
          <fields>
            <field name="alertID" type="int64" key="true"/>
            <field name="pt_id" type="int64"/>
            <field name="GPS_longitude" type="double"/>
            <field name="GPS_latitude" type="double"/>
            <field name="speed" type="double"/>
            <field name="course" type="double"/>
            <field name="time" type="stamp"/>
            <field name="poly_id" type="int64"/>
            <field name="poly_desc" type="string"/>
            <field name="poly_dist" type="double"/>
            <field name="event_nb" type="int64"/>
            <field name="POI_id" type="int64"/>
            <field name="POI_desc" type="string"/>
            <field name="POI_dist" type="double"/>
            <field name="vehicle" type="string"/>
            <field name="alert" type="string"/>
          </fields>
        </schema>
        <patterns>
          <pattern name="enter_polygon">
            <events>
              <event source="geofence_circles"
                name="e0">isblank(poly_id)</event>
              <event source="geofence_circles" name="e1">not
                isblank(poly_id)</event>
            </events>
            <logic>fby{5 seconds}(e0, is(e1))</logic>
            <output>
              <field-expr node="e1">pt_id</field-expr>
              <field-expr node="e1">GPS_longitude</field-expr>
              <field-expr node="e1">GPS_latitude</field-expr>
              <field-expr node="e1">speed</field-expr>
              <field-expr node="e1">course</field-expr>
              <field-expr node="e1">time</field-expr>
              <field-expr node="e1">poly_id</field-expr>
              <field-expr node="e1">poly_desc</field-expr>
              <field-expr node="e1">poly_dist</field-expr>
              <field-expr node="e1">event_nb</field-expr>
              <field-expr node="e1">POI_id</field-expr>
```

```
              <field-expr node="e1">POI_desc</field-expr>
              <field-expr node="e1">POI_dist</field-expr>
              <field-expr node="e1">vehicle</field-expr>
              <field-expr node="e1">
                <![CDATA[ vehicle & " has entered " & poly_desc ]]>
              </field-expr>
            </output>
            <timefields>
              <timefield source="geofence_circles" field="time"/>
            </timefields>
          </pattern>
          <pattern name="exit_polygon">
            <events>
              <event source="geofence_circles"
               name="e0">isblank(poly_id)</event>
              <event source="geofence_circles" name="e1">not
               isblank(poly_id)</event>
            </events>
            <logic>fby{5 seconds}(e1, is(e0))</logic>
            <output>
              <field-expr node="e1">pt_id</field-expr>
              <field-expr node="e1">GPS_longitude</field-expr>
              <field-expr node="e1">GPS_latitude</field-expr>
              <field-expr node="e1">speed</field-expr>
              <field-expr node="e1">course</field-expr>
              <field-expr node="e1">time</field-expr>
              <field-expr node="e1">poly_id</field-expr>
              <field-expr node="e1">poly_desc</field-expr>
              <field-expr node="e1">poly_dist</field-expr>
              <field-expr node="e1">event_nb</field-expr>
              <field-expr node="e1">POI_id</field-expr>
              <field-expr node="e1">POI_desc</field-expr>
              <field-expr node="e1">POI_dist</field-expr>
              <field-expr node="e1">vehicle</field-expr>
              <field-expr node="e1">
                <![CDATA[ vehicle & " has exited " & poly_desc ]]>
              </field-expr>
            </output>
            <timefields>
              <timefield source="geofence_circles" field="time"/>
            </timefields>
          </pattern>
        </patterns>
      </window-pattern>
```

The window geofence_circles also feeds a Filter window named filter_null_POI.

Example Code 7.19

```
<window-filter name="filter_null_POI" collapse-updates="true"
index="pi_HASH">
        <expression>
          <![CDATA[ not isnull(POI_id) ]]>
        </expression>
</window-filter>
```

The window pattern_alerts_polygons feeds a Remove State window named
alerts_polygon_inserts. It converts events that it receives into Inserts and adds a field named

eventNumber, which is a monotone-increasing integer. This is the key field of the Remove State window. This window filters out any Delete events with the retention flag set.

Example Code 7.20

```
<window-remove-state name="alerts_polygons_inserts"
remove="retentionDeletes"/>
```

The window filter_null_POI feeds a Copy window named copy_alert_points. Like the other Copy windows, this sets a retention policy of bytime_sliding.

Example Code 7.21

```
<window-copy name="copy_alert_points" collapse-updates="true"
index="pi_HASH">
        <retention type="bytime_sliding">1 seconds</retention>
</window-copy>
```

This copy window feeds two Aggregate windows. The first is named alerts_points.

Example Code 7.22

```
<window-aggregate name="alerts_points" collapse-updates="true"
index="pi_HASH">
        <schema>
          <fields>
            <field name="POI_id" type="int64" key="true"/>
            <field name="vehicle" type="string"/>
            <field name="pt_id" type="int64"/>
            <field name="GPS_longitude" type="double"/>
            <field name="GPS_latitude" type="double"/>
            <field name="time" type="stamp"/>
            <field name="POI_desc" type="string"/>
            <field name="POI_dist" type="double"/>
          </fields>
        </schema>
        <output>
          <field-expr>
            <![CDATA[ ESP_aFirst(vehicle) ]]>
          </field-expr>
          <field-expr>
            <![CDATA[ ESP_aLast(pt_id) ]]>
          </field-expr>
          <field-expr>
            <![CDATA[ ESP_aLast(GPS_longitude) ]]>
          </field-expr>
          <field-expr>
            <![CDATA[ ESP_aLast(GPS_latitude) ]]>
          </field-expr>
          <field-expr>
            <![CDATA[ ESP_aLast(time) ]]>
          </field-expr>
          <field-expr>
            <![CDATA[ ESP_aLast(POI_desc) ]]>
          </field-expr>
          <field-expr>
            <![CDATA[ ESP_aLast(POI_dist) ]]>
          </field-expr>
```

```
        </output>
</window-aggregate>
```

The second Aggregate window is named alerts_vehicles. It performs calculations with regard to the vehicle.

Example Code 7.23

```
<window-aggregate name="alerts_vehicles" collapse-updates="true"
index="pi_HASH">
        <schema>
          <fields>
            <field name="vehicle" type="string" key="true"/>
            <field name="POI_desc" type="string" key="true"/>
            <field name="GPS_longitude" type="double"/>
            <field name="GPS_latitude" type="double"/>
            <field name="POI_dist" type="double"/>
            <field name="time" type="stamp"/>
          </fields>
        </schema>
        <output>
          <field-expr>
            <![CDATA[ ESP_aLast(GPS_longitude) ]]>
          </field-expr>
          <field-expr>
            <![CDATA[ ESP_aLast(GPS_latitude) ]]>
          </field-expr>
          <field-expr>
            <![CDATA[ ESP_aLast(POI_dist) ]]>
          </field-expr>
          <field-expr>
            <![CDATA[ ESP_aLast(time) ]]>
          </field-expr>
        </output>
</window-aggregate>
```

Create Connections to Collect Data

Now create connections to collect the data that is produced by selected windows of the streaming project. Use the getEventCollection method to gather events from specified windows of the continuous query.

Example Code 7.24

```
In [ ]:
tracker = conn.getEventCollection("p/cq1/tracker")
circles = conn.getEventCollection("p/cq1/circles_in")
polygons = conn.getEventCollection("p/cq1/poly_in")
poiAlerts = conn.getEventCollection("p/cq1/alerts_vehicles")
polygonAlerts = conn.getEventStream("p/cq1/alerts_polygons_inserts")
```

Create a Map

Now, create the map of Paris. You add specific circles and polygons to this map using paris.addCircles and paris.addPolygons. These circles and polygons act as a boundary for your geofence.

Create a Map

Now, create the map of Paris. You add specific circles and polygons to this map using paris.addCircles and paris.addPolygons. These circles and polygons act as a boundary for your geofence.

Example Code 7.25

```
In [ ]:
colors = ["#e03531","#ff684c","#f0bd27","#51b364","#8ace7e"]

paris =
visuals.createMap(tracker,lat="GPS_latitude",lon="GPS_longitude",si
ze=10,color="speed",colors=colors,color_range=(0,50),

title="Paris",popup=["vehicle","speed"],marker_border=False,
                zoom=15,tracking=True,center=(48.875,2.287583))

paris.addCircles(circles,lat="POI_y",lon="POI_x",radius="POI_radius
",text="POI_desc")
paris.addPolygons(polygons,coords="poly_data",text="poly_desc",order="
lon_lat")
```

Create a Display for the Geofence

Next, you create a table and chart that displays information such as when the car leaves the circles and polygons that you created in the previous step.

Example Code 7.26

```
In [ ]:
poiAlertsChart =
visuals.createBarChart(poiAlerts,y=["POI_dist"],title="Point of
Interest Alerts Distance",

width="60%",orientation="horizontal",xrange=(0,100))
polygonAlertsTable =
visuals.createTable(polygonAlerts,values=["_timestamp","alert"],tit
le="Polygon Alerts",
                            width="40%",reversed=True)
```

Create a compass and a speedometer. The compass indicates which direction the car is heading, while the speedometer indicates the vehicle's speed.

Example Code 7.27

```
In [ ]:

heading =
visuals.createCompass(tracker,heading="course",size=300,columns=4,

reciprocal_color="#e8e8e8",heading_color="#89cff0",outer_color="#89
cff0",
                            bg_color="#f8f8f8",line_width=2)
```

```
speed =
visuals.createGauge(tracker,value="speed",size=300,columns=4,colors
=colors,shape=80,
                            segments=5,label="Speed",range=(0,40))
```

Define a Publisher

You define a function that requires user input for the vehicles' name. This block of code produces a widget that requires you to enter a vehicle name of your choice.

You must run **Code Block A** and **Code Block B** before entering a vehicle name.

After the map and charts are populated, you enter a vehicle name of your choice and click **Publish**. You can enter up to two vehicles at a time.

Example Code 7.28

```
In [ ]:
# Code Block A

def publish(b):
    MyPublisher(vehicleName.value)

vehicleName = widgets.Text(description="Vehicle",value="Car
1",layout=widgets.Layout(width="30%"))
button = widgets.Button(description="Publish")
button.on_click(publish)

box = widgets.HBox([vehicleName,button])
gauges = widgets.HBox([speed,heading])
box
```

Running this code produces the following output.

Figure 7.2: Widget Produced by Example Code 7.28

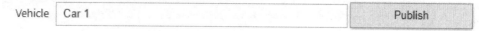

Running the following code produces a blank map of Paris and a blank graph for "Points of Interest Alerts Distance."

Example Code 7.29

```
In [ ]:
# Code Block B

alertsBox =
widgets.HBox([poiAlertsChart,polygonAlertsTable],layout=widgets.Lay
out(height="350px"))
widgets.VBox([paris,gauges,alertsBox])
```

Figure 7.3: Widget Produced by Example Code 7.29

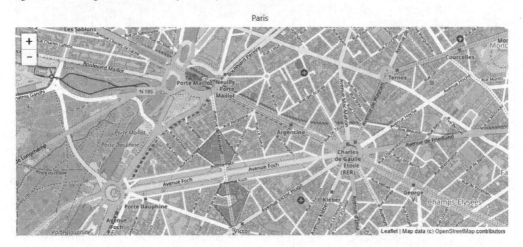

Now suppose that you retain the default car name "Car 1" and click **Publish**. The graph becomes animated. A dot that represents a car travels the streets, occasionally crossing the circle geofence and the polygon geofence.

Figure 7.4: Animated Widget

Simultaneously, the compass and speedometer indicate the direction and speed of the car.

Figure 7.5: Compass Shows Changes in Direction

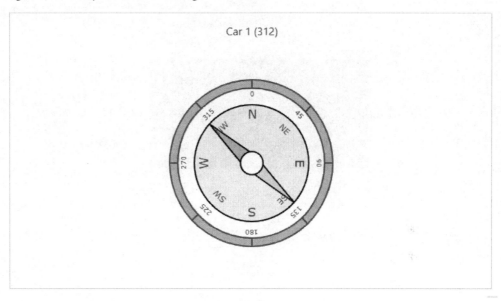

Figure 7.6: Speedometer Shows Changes in Speed

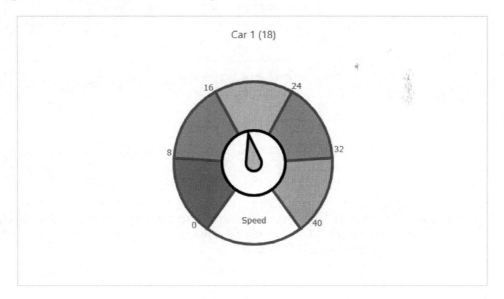

The graph shows continuously updated data about the state of the car.

Figure 7.7: Graph Displays Updated Data About Car

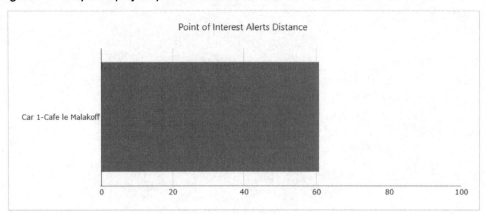

Figure 7.8: Table Updates with Alerts

Polygon Alerts

_timestamp	alert
2020-01-07 18:43:25	Car 1 has exited polygon 3
2020-01-07 18:43:21	Car 1 has entered polygon 3
2020-01-07 18:43:16	Car 1 has exited polygon 3
2020-01-07 18:43:11	Car 1 has entered polygon 3
2020-01-07 18:43:10	Car 1 has exited polygon 2
2020-01-07 18:43:05	Car 1 has entered polygon 2

Suppose you return to the Vehicle box that was produced when you executed Code Block A and insert the name "Car 2." Then you click **Publish**. Now, two vehicles are traversing the streets. Each vehicle is represented by a different dot.

Figure 7.9: Two Cars Tracked in Animated Graphic

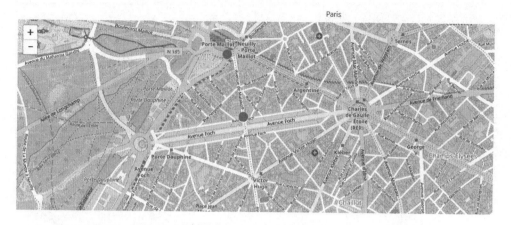

The compass and speedometer now track both vehicles.

Figure 7.10: Two Cars Tracked by Compasses

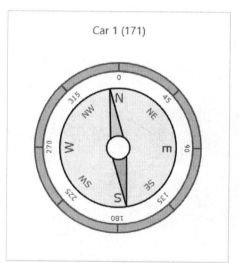

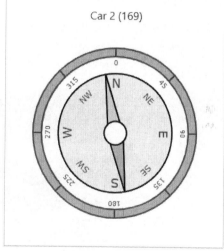

Figure 7.11: Two Cars Tracked by Speedometers

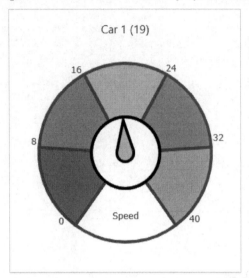 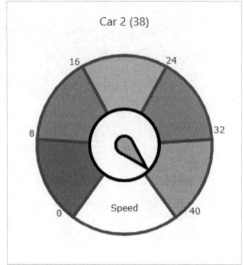

The graph provides information about both vehicles.

Figure 7.12: Graph Displays Updated Data About Two Cars

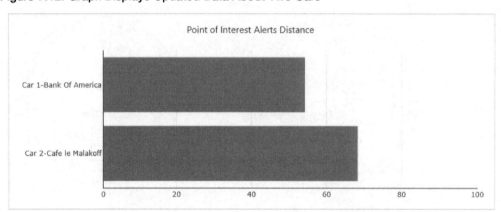

The table tracks both vehicles with respect to the polygon and circle geofences.

Figure 7.13: Table Updates with Alerts from Two Cars

Polygon Alerts

_timestamp	alert
2020-01-10 14:08:58	Car 1 has entered polygon 2
2020-01-10 14:08:55	Car 1 has exited polygon 2
2020-01-10 14:08:53	Car 1 has entered polygon 2
2020-01-10 14:08:50	Car 1 has exited polygon 1
2020-01-10 14:08:44	Car 1 has entered polygon 1
2020-01-10 14:07:26	Car 2 has exited polygon 3

Conclusion

The Geofence window included with SAS Event Stream Processing window provides capabilities for processing geolocation data in real time. It applies intelligence at the edge by analyzing movements and locations of people or connected objects. It opens the horizon to new IoT applications in countless domains in order to react immediately and appropriately. Should a child leave a particular location, a parent can be immediately notified. If a vehicle strays from its garage, then its owner can be alerted right away. If a tourist wanders from a designated area within a park or museum, security can respond within seconds. The only limit to the application of a Geofence window is the user's imagination.

Reference

White, S. K. (2019). "What is geofencing? Putting location to work," https://www.cio.com/article/2383123/geofencing-explained.html

About the Contributor

As Head of IoT Support, Enablement and Innovation for EMEA and AP for SAS, **Frédéric Combaneyre** supervises the real-time data streaming architecture for multi-national customers across a wide cross-section of industries, helping organizations develop Internet of Things projects and initiatives. He actively participates in SAS streaming analytics solutions design and expansion, working closely with the SAS R&D organization.

Acknowledgments

The authors would like to acknowledge Robert Levey for developing the visualization programing objects used in the example and Carlos Sebastiani for implementing the example in the SAS Analytics Cloud.

Chapter 8: Using Deep Learning with Your IoT Digital Twin

By Brad Klenz

Introduction

As the Internet of Things (IoT) grows, connected devices are often located in remote places, operating in a variety of physical environments. These devices communicate with control systems and with each other. It can be challenging to monitor whether devices in these disparate environments are operating properly and efficiently. To meet these challenges, you can create a digital twin of a device. A *digital twin* is a virtual representation of the device in real time. It can tell you how a device is operating, regardless of the device's environment.

Sensors are installed on IoT devices to monitor their health and the environment around them. Analytics can be applied to this sensor data in order to create a true real-time digital twin. With analytics, a digital twin can do the following:

- Fill in gaps where sensor data is not available.
- Notify when a device is not operating efficiently.
- Provide advance notice when a device is failing.
- Detect when devices are not interacting properly.
- Forecast future operating conditions.

More recently, deep learning and artificial intelligence are being applied to data-rich applications like digital twins. These methods can be added to your digital twin for more understanding of the remote assets. Image and video analytics can be used to capture operating conditions that are missed by regular sensors. Recurrent neural networks (RNN) add temporal data analysis and pattern detection in real-time data streams that are prevalent in digital twins. Reinforcement learning can be applied to control an asset and achieve optimal output.

How Can Analytics Be Used to Create a Digital Twin?

Without analytics, a digital twin simply reports data that is collected remotely. To know the temperature of a building in another city, you read a temperature sensor and report a value on a dashboard. You can also save measurements in a database and review previous values. You can summarize data and create reports with the average temperature and the observed range of temperature values. This information can provide a sense of the general status of the building.

To obtain a comprehensive view of a remote device, you must be able to accommodate missing data. You must be able to handle abnormal readings. You should develop a model to predict future states of the sensors. And you should be able to apply insights gathered from multiple sensors to improve the data collection capacity of each one. With analytics applied to sensor data, you can fulfill these objectives. Table 8.1 explains how analytics can help.

Table 8.1: How Analytics Can Help Obtain a Comprehensive View of a Remote Device

Objective	How Analytics Can Help
Filling in the Gaps	Even with an abundance of sensor data, there might be times when something important is not measured. Sensors could be down. Something could happen during a time interval that the sensor does not support. Filling in these gaps with analytics is necessary to gain a full understanding of the device's operation.

Objective	How Analytics Can Help
Anomaly Detection	A real-time dashboard of sensor data can report the current status of a device. A more interesting view would be to show when and where a device is behaving abnormally and indicate what is different. What is defined as normal operation can depend on many factors. The device could have different configurations, it might operate differently due to expected changes in the surrounding environment, or it might have a known degradation pattern.
Looking Forward	A database of previous sensor measurements can be used to review all previous operating states of a device. In some cases, though, you might want to know the future operating state or determine the operating state under a set of conditions that have not been seen before. Using an analytic model of the device can estimate the results under these unknown conditions.
Learning from All Your Devices	The same type of device can be installed in many places. Over time, new devices are installed. Each of these devices might have unique characteristics that relate to its usage and environment, but you can learn something from each device and apply that knowledge across devices. When building analytic models of our devices, the model errors can tell us what characteristics are well understood and what characteristics need further research.

Digital Twin Examples

Smart Grid

The existing power grid is a well-understood system with much historical data and analysis under many operating conditions. The addition of new distributed energy resources (DER) like renewable (solar, wind) generation and energy storage is changing how the power grid is operated. These new generation sources are less reliable than historical power sources. Solar and wind generation produce power only when solar and wind resources are available. A solar farm's output drops dramatically with each cloud that passes by (Figure 8.1). This is a localized condition that might apply to a remote location with only basic weather data available. These new power sources are being incorporated into a system that has a very high standard of reliability, the power grid on which we all depend. New sensors model the power grid in much greater detail, including more detail about load conditions.

Figure 8.1: Solar Farm

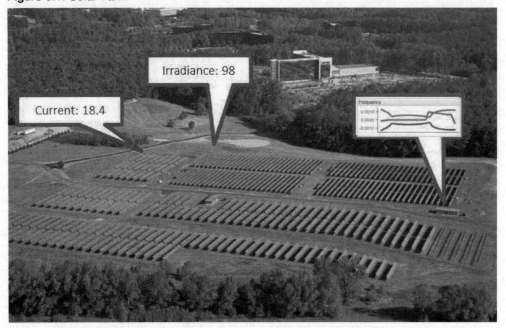

Connected Vehicle

Modern cars and trucks have approximately 400 different sensors taking measurements (Figure 8.2). This includes mechanical measurements such as temperatures and pressures of the main systems (engine, transmission, brakes). It also includes emissions measurements and operating conditions such as the number of passengers or environmental data. One of the most challenging aspects of connected vehicle data is the need to run the analytics on the vehicle. Although there is much data available, the cost to send the data from the vehicle is prohibitive.

Figure 8.2: Connected Truck

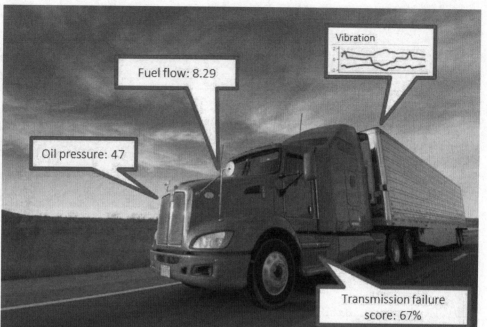

Smart Building

Buildings consume a lot of energy. Recently constructed buildings often have sensors to support their automation systems (Figure 8.3). These sensors capture data on the state of systems such as HVAC, water, and lighting, as well as their control systems. Additional data can be added to capture the environment (weather), occupancy, and usage conditions. The objective is to operate the building at a comfortable level for the occupants in an energy-efficient manner. In addition, equipment in the building can be monitored for potential failure or performance degradation. Analytic models are used to determine when a part is likely to fail so that it can then be serviced ahead of time, avoiding an unplanned outage.

Figure 8.3: Smart Building Sensors

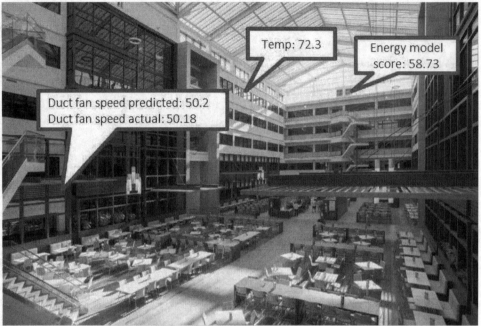

Sensors Might Be Too Expensive

Just because there is a sensor available in the market to capture a desired measurement, it does not mean that it is a good business decision to install it. In the smart grid use case, one of the available sensors is called a phasor measurement unit (PMU). These sensors cost about $2,500 each, with additional cost for the installation and network communications. This cost is justifiable for installation at power generation facilities and larger substations. But until technology costs drop, PMUs are not justifiable at something like the household or building level.

Even though it is cost-prohibitive to install an advanced sensor on every device, the expense is justifiable for some devices. Selectively installing advanced sensors serves a couple of purposes. First, you will start collecting detailed data at selected locations. This data is then used to model these selected devices, and you can apply that model to devices that lack the advanced sensors. Second, the detailed measurements from some selected locations can be used to build a more detailed model of the entire system. In the power grid use case, there are relationships with power generation and load across the grid. Having the advanced sensors in some locations captures important details about these interactions.

Sensors Might Interfere with the Device

Although a sensor might be available to capture a measurement, having the sensor installed can affect the operation of the device. One example is an airflow sensor. In a combustion turbine, airflow is critical to the efficient operation of the engine. Unfortunately, the presence of the sensor in the airflow alters the function of the turbine. Other examples of sensor interference include cases where the weight of the sensor affects the operation or balance of the device.

Sensors Might Not Communicate Well

You might find situations where communication with the sensor presents challenges. With devices in remote locations, high-speed, reliable network bandwidth might not be available or costs too much to be practical. In addition, devices that are in motion can also have sporadic connections. In connected vehicle applications, there is a large amount of data on the vehicle being generated at high frequency. However, the cellular connection costs to do the data analysis off the vehicle are not economical. In this case, we run the analytic model on the vehicle, with the full fidelity of the data available. We use the analytic model to determine the key characteristics of the vehicle's operation and send the selected results over the cellular connection. A larger system model looking at vehicle traffic flows would then benefit from the detailed data captured on the vehicle and have a cost-effective means to model the traffic flows.

Data Might Be Collected at Different Intervals

Some sensors are designed to collect data at different time intervals. This might be a simple way to reduce the cost of the sensor. The measurement being taken might be known to slowly change values, and a longer interval reduces the network traffic requirements for the system. In a smart building application, some equipment is monitored closely for vibration at higher data rates (for example, chillers). Other measurements change more slowly (for example, building air temperatures). Another example of data that are sampled at different time intervals is when various sensor technologies have been installed over time. Older sensors can have a slower data rate than newer sensors.

Analytic Techniques to Fill the Gaps

It is helpful to think of a physical system as a fabric or surface over which many measurements and relationships flow. This involves thinking about the physical system both spatially and temporally. As discussed above, there are many reasons why we have data measurements only at specific points on this surface, even though we know that the true physical system is continuous.

For the points in space and time where we have gaps, we create an analytic model to determine the relationships. These analytic models use predictive model algorithms like regression, neural networks, or gradient boosting.

In one smart building project at SAS, there are more than 100 separate analytic models to model the HVAC airflow within the building.

Anomaly Detection

A real-time system model for your digital twins often tells you the system is operating normally. It would be more useful for the system model to indicate when the system is not operating normally. This knowledge directs attention to places where the system could be improved. The concept of looking for times when the system is not operating normally is called *anomaly detection*.

As you start to look at the data, you will see that many different conditions could be called anomalous. One example is when your system model is estimating conditions in the system or system output, but you are not seeing those conditions in real time. Another example of an anomaly is when you have determined that the system is operating in a way that should not be possible. Yet another type of anomaly involves looking at the change of state over time. You might see a pattern of states where each state is in normal conditions, but the pattern of the states over time is not normal. An abnormal pattern is another type of anomaly.

One important aspect of our definition of an anomaly is that an anomaly does not necessarily signal a problem. All we are saying is that the building is not operating as expected. The causes could be some state that you have not previously seen before, or it could be an equipment issue.

As your system model improves over time, it reflects a better understanding of the system. This results in flagging fewer and fewer situations as anomalies because they are now explained phenomena.

Using Your System Model for Anomaly Detection

Let us first look at using your system model to detect anomalous behavior. With your system model, you can incorporate many different inputs from your system. This model is targeted toward higher-level outcomes. For example, in our smart building application, we can build the system model to estimate energy consumed or maintain a comfort level based on temperature, humidity, and air mixture. The system model would use all the different inputs that we have available. With smart buildings, energy sensors are mounted on various pieces of equipment. Some of these sensors monitor the air flows on the HVAC system, and some incorporate environmental conditions, such as the weather, temperature, and humidity. There are also input variables to capture the building usage at a given moment.

Your system model uses those inputs to estimate the higher-level target variable, such as energy usage. With the system model now deployed on the building, you can monitor all those input variables and estimate the target variable. Then you compare the estimate to some actual measurement of that variable, such as the power meter coming into the building for energy usage. If the target variable is temperature from an inside air sensor, you compare the actual measurement from the sensor to the estimate from the system model. In many cases, this estimate is very close to the actual measurement of the target variable. What you are looking for is cases where the estimate is off. In these cases, the system is not reacting as it previously has to this set of conditions.

Note that this model is continuous, so the estimate is off by a factor. Still, the model gives us a set of confidence limits, which you can use to determine when you actually send a notification or alert or flag an anomaly condition. We could choose the warning limits, based on previous experience, around the level of 90% or 95% confidence. This yields an initial set of notifications.

At this point you have notifications, but the next step will be to determine the root cause of the anomaly. Because you are monitoring all the input variables in real time, you can snapshot the state of the system when the anomaly occurred. You can even snapshot a time window before and after the anomaly to get more context. You can use the snapshots to then do root-cause analysis to identify what condition triggered the anomaly.

The algorithms to use for the system model are predictive model algorithms such as neural networks. You should use these models due to the large number of input variables that you are modeling. These system models are also deep-learning models, which can handle very large numbers of interactions among the input variables.

Now that you have some very high-level, or macro-level, system models that are monitoring the system, you can further refine these models. You can add more specific predictive models for individual systems within the overall system. The smart building application uses predictive models for many of the elements in the HVAC system. These are elements such as the air handler units (AHU), the fan power boxes, and the variable air volume (VAV) boxes.

By creating these additional device-level models, you get more fine-grained granularity for determining where an anomaly occurs. You also capture anomalies that might be lost in the larger variations of the overall system. Figure 8.4 depicts the use of SAS Report Viewer to inspect the differences between modeled and actual energy usage.

Figure 8.4: System Model Estimate Versus Actual Usage

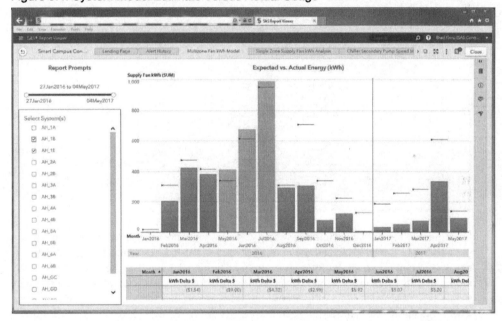

Another example of using the system model is on our power grid application where we are looking for situations that threaten grid stability. In this scenario, we have modeled various grid sensors like voltage and current, as well as the operating state of equipment such as static VAR compensators. The system model then detects and alerts us when we are in a situation where we are vulnerable to a grid collapse.

Operating Modes for Anomaly Detection

Now consider a different type of anomaly detection. In some cases, your historical data can determine how many different operating modes your system assumes. A good example is energy usage within our smart building application. To capture the complexity of energy usage, you can place sensors on the different subsystems of energy consumption in the electrical system in the building. You can also have sensors on the HVAC and subcomponents. You can have sensors on the lighting. You can have sensors on the office computer systems and on the non-critical plug loads like the break rooms and appliances.

To perform this type of anomaly detection, use an algorithm like Support Vector Data Description (SVDD). This algorithm uses a distance-based method based on clustering. It works across many different dimensions simultaneously and determines a binary categorization. You train the model using data from known good operating modes. The model determines how much the different dimensions contribute to that mode. After you have built a model, you can then use the model in real time to monitor the building. When monitoring the building, you take all the input variables and their current mode and compare it to your model. The model detects any combination of states that you have not seen before. This situation is then flagged as an anomaly. Figure 8.5 shows the application of the SVDD model to sensor data to detect anomalies.

Figure 8.5: Support Vector Data Description Model and Degradation

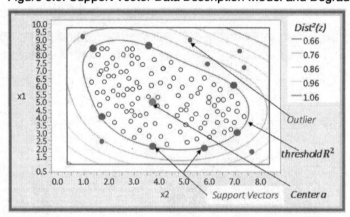

In implementing our smart building energy model, we were able to see a relationship between the lighting in the building and the computer usage. This makes sense in that typically employees come into the building, turn on their lights, and then use their computer. The model detects these measurements as ones that are correlated. An example of an anomaly would be a case where a lot of the building lights are on, but there is very little computer usage. This indicates a mode not seen before. As discussed earlier, these anomalies might not necessarily be problems, but just unusual operating conditions. In the example where the lights are on, but computer usage is low, it is possible that building maintenance is occurring.

The engineers doing building maintenance turned on the lights to be able to do their work, but they are not using the computers within the offices. This would be an unusual condition, but it would not necessarily indicate a problem.

Changes in Relationships Between the Parts of Your System

Your digital twin might have relationships between parts of your system that can help detect a different type of anomaly. One case is when you have similar machines that are operating in a similar environment, and it is expected that these machines would have similar performance in those environments. Examples would be a power plant where multiple generators are working in parallel or multiple chillers in a building HVAC system. A second case is where your digital twin has a set of similar parts that function in parallel. An example would be individual cylinder measurements in a combustion engine. These represent more complex problems than a part that always operates in a controlled environment.

For anomaly detection, you use an algorithm called Moving Window Principal Components Analysis (MWPCA). This method starts with computing the principal components from the measurements of the related parts. It takes the correlated measurements and creates linearly uncorrelated values that are the principal components.

These principal components capture most of the variability in the measurements. These are the environmental effects. In our power generation example, the variability is the fluctuating power load on the generators. For the building's HVAC chillers, it is the temperatures of the coolant.

Then we add a moving window to account for the time dimension of the problem. As the environment changes over time, the principal components for the monitored parts should change in a related manner. A change in the principal components would then indicate changes to the relationships among the parts, after accounting for the changes in the environment. A comparison of the environment changes to the changes in principal components is depicted in Figure 8.6.

Figure 8.6: Measurements with Environmental Effects Compared with Principal Components

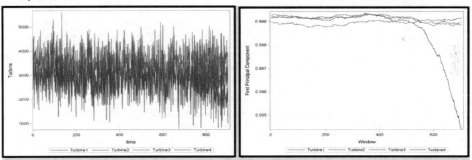

Another benefit of this algorithm is that you do not necessarily need measurements of the environmental conditions. If you know from the system design that the parts operate in a similar environment, the part measurements suffice to detect anomalies.

The algorithm also makes it easy to determine the part of the system that is causing the anomaly. The loading value for the principal components point to the measurements for the bad part.

Changes in Patterns Over Time

When collecting measurements for your digital twin, many measurements are time series of individual values taken from a sensor. Examples mentioned earlier include duct air pressure in an HVAC system or power consumed by an HVAC component like a chiller. Looking at these measurements over the time series, you will see regular patterns in the data when the system is operating normally. When an anomaly occurs, it might show itself as a different pattern in the time series that has not been seen before, or that is a rare occurrence.

We can detect this type of anomaly using an algorithm called motif analysis. In motif analysis, the patterns are also called subsequences. The method analyzes the time series and starts to identify small, repeating subsequences. These small subsequences might repeat over and over during normal operation, which can then identify larger subsequences.

An anomaly is then detected by ranking the subsequences that are identified. Normal operations are characterized by long subsequences or a subsequence that is seen repeatedly with a short interval between occurrences. An anomaly subsequence is unique within the time series or has a much longer interval between occurrences, as shown in Figure 8.7.

Figure 8.7: Time Series with Anomaly Subsequences Identified

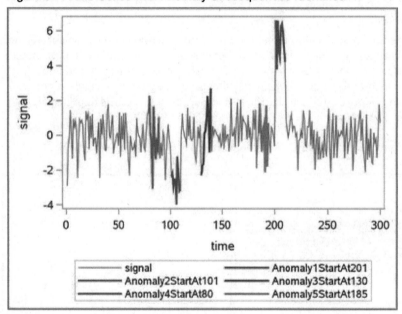

Note that motif analysis is done on univariate time series. The input time series could be the result of a multivariate model that captures a relationship that you want to analyze.

Predicting the Future with Your Digital Twin Model

Until this point, the analytics model of your digital twin monitored your IoT devices in real time. You were able to fill in gaps where measurements were lacking. You were able to detect anomalies when the system was not behaving normally. Another important use for your analytics model is to look beyond the real-time, current state of your IoT devices. You can also use your analytic model to predict different states from the current state, to forecast a future state of a device, and to simulate the device in large numbers of possible conditions.

Using your analytic model to predict different states for your device, you now can perform "what-if" analysis. In this case, you feed your model different inputs for the state that you are interested in. The model then gives you the output state based on those inputs. This enables you to check different conditions that you might have or that you think you might encounter. This enables you to learn more about your device using the digital twin. Note that these states are not just going back in history and finding a comparable time. You can look for new combinations of state that you have not seen directly before.

Another example of using an analytics model to look beyond real-time monitoring is to use your model to forecast future states. In this case, you can obtain estimates of various input variables that are anticipated for your model in the future. For example, in the smart building application, you can look at the expected usage and occupancy of the building at a future date. You can also get the weather forecast for that future date. Then, using the system model, you can forecast output results, such as energy usage or comfort level.

Using Your Digital Twin Model for Simulations

The third example of using your analytic model for non-real-time monitoring is to perform simulations of the device. Device simulations let you examine a range of possible input values and understand the resulting outputs. The simulations are useful for understanding the range of inputs that can be seen.

Simulations also help you understand various congestion points, or stress points, on the system. The simulations are most helpful in cases where you have built subsystem-level models.

The simulations enable you to understand the results at various points in your overall system using the subsystem-level models. By running the simulation over a wide range of possible input states, you can see the output state for many conditions that have not previously been observed. Your model captures interaction effects between observable states and combinations of states where the interactions cause undesirable results. Thus, running the simulation lets you identify undesirable situations and explore them further. Examining these undesirable situations, you can review your historical data and see whether they have been encountered in the past. You can gain a level of insight into how likely it is that you will see this undesirable situation. The simulation also gives you information to identify shortcomings in your system. Knowing those shortcomings, you can improve the system specifically to remediate known problem areas.

Building Your Digital Twin Model

Now that you understand what your digital twin model is and how you use it, let us look how to build it. To build your digital twin model, you need to add to your knowledge in two areas. First, you need to understand the analytics life cycle, historical data, and streaming data. Second, you need to understand the physical and software environment that needs to be constructed for the digital twin model.

One tip for building the system is to construct a learning loop. The primary objective at first is to get an end-to-end system working and then incrementally add to this system.

The analytics life cycle (Figure 8.8) provides the framework to create your digital twin models, improve these models, and manage the models.

Figure 8.8: IoT Analytics Life Cycle

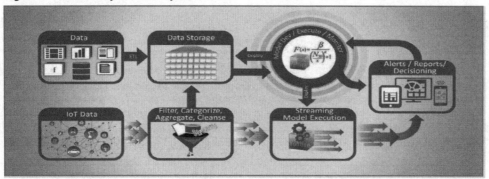

The analytics life cycle takes your streaming data that is coming from your sensors, manages this data in real time, and provides the data for your historical context as needed. Even without real-time streaming analytics, many companies start by taking sensor data and collecting it in a large database. You can use this database to build initial analytic models that describe your digital twin. You can take recurring data in a batch mode and process it to achieve some of your objectives for the digital twin. You can find normal states, summarize observed states, and revise the analytic model as you collect more data to improve the model. If you observe the model over time, you will generate ideas on how to improve your physical system.

To realize the full benefits of your analytic model for your digital twin, you do need to incorporate the model into a real-time streaming environment. The real-time streaming environment provides the full-time, visible, digital equivalent of your physical systems. The streaming environment must be able to capture the sensor data that you need and implement the numerous analytic models that you have for your digital twin. One benefit of your digital twin is to provide real-time status of your remote devices. Therefore, the streaming environment must also be able to integrate with control environments and real-time dashboard systems.

There is, of course, a physical and software environment required to support your digital twin in the physical environment. This environment is called an edge-to-enterprise system. In your physical environment, you must have sensors on the remote devices. The sensors then feed into edge computing gateways to provide analytics right at the source of the sensor data. From your edge computing and remote devices, you then use network communications back to a central on-premises server or cloud environment. Having a cloud environment enables you to create analytic models that include many devices. Figure 8.9 shows how your IoT system can comprise multiple levels.

Figure 8.9: Edge to Cloud Architecture

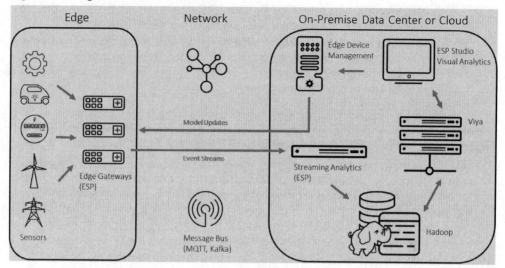

Applying Deep Learning Techniques

Analytics provide the means to monitor a device's state, predict failure, and detect anomalies. You can use *deep learning models* on your digital twin to obtain an even deeper understanding of your IoT system.

Image and video analytics can be used to capture operating conditions that are missed by regular sensors. Recurrent neural networks (RNN) add temporal data analysis and pattern detection in real-time data streams that are prevalent in digital twins. With these deep learning capabilities, your digital twin provides a new level of insight for your remote devices.

Deep learning is becoming more prevalent, with some typical use cases emerging. Natural language processing (NLP) is used with personal assistants and chatbots. Computer vision is used for object recognition in autonomous driving and detecting tumors on medical images. Facial recognition is used for access control. Reinforcement learning is being researched in game applications.

To add more value to your digital twin, deep learning can be added for specific use cases. These use cases are targeted based on typical sources of data found in digital twin applications.

The deep learning techniques that you can use include the following:

- **Computer vision.** The use of images and video to provide insight not easily available with existing sensors. Cameras can sometimes be installed much more easily than other sensors. Cameras can also be retrofitted to existing assets passively, while adding sensors can be more invasive.

- **Recurrent neural networks (RNN).** Sensor data frequently captures measurement over time, and we are looking for issues that develop over time. RNNs can provide complex pattern recognition as well as specialized forecasting.
- **Reinforcement learning (RL).** Your digital twin is frequently a twin of a physical asset that is being controlled for optimal operation. The output of the physical asset is typically captured. Using RL, we can learn how the controls of the asset can be set to learn how to achieve optimal output.

Real-time Application of Deep Learning in Your Digital Twin

Deep learning requires a lot of computing power. The deep learning models are trained on large databases and are almost always done offline. It is not unusual to take hours or days to train a model. Once the model is trained, the application of the model through inferencing requires more computing resources than typically required for digital twin applications. For some applications, you can use near real-time or slightly delayed results. For example, in the computer vision defect detection described in the next section, it might be acceptable to hold a production batch while the defect detection is performed. In other cases, real-time inferencing is needed. Inferencing can be done in the cloud or in a data center where resources are readily available. For edge inferencing, edge gateways are now becoming available with ample compute power, but you must plan for this specialized need (McGrath 2018).

Applying Computer Vision Techniques

Popular uses of computer vision include facial recognition and object detection. To see where computer vision can help a digital twin, look for applications that would require visual inspection. Some examples are given in the paragraphs that follow.

Defect detection in semiconductor manufacturing. In semiconductor manufacturing of wafers and dies, many tests cannot be run until the packaging phase. With computer vision, it is practical to take images of the wafer earlier in the production process and inspect it for issues (Figure 8.10). The inspection can be used to find defects and determine the number and location. This allows an earlier determination of final yield from the wafer. With previous labeled images of diagnosed defects, it is also possible to classify the defect types using computer vision. This helps augment a larger root cause analysis for process improvement.

Figure 8.10: Silicon Wafer and Wafer Defects

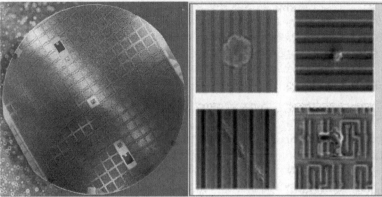

Defect detection in discrete parts. With visual inspection of discrete parts, you can more easily catch production defects. These are various issues in production quality. For example, in aerospace and automotive parts production, you can determine incomplete finishes or casting issues (Figure 8.11).

Figure 8.11: Automotive Engine and Casting Defects

Infrared patterns for heat buildup in power substations. Using specialized cameras, you can capture images for different spectrums. Infrared cameras can capture a more complete picture of temperature deviations and patterns than what would be possible with individual temperature sensors. This allows for new applications such as monitoring power substations for components getting ready to fail.

Implementing a Computer Vision Model

The process for implementing a computer vision model is as follows:

- If possible, fix the camera to a stable mount point so that all images are taken from the same angle and with the same proportions. This vastly simplifies the model training as compared to general object recognition models, which must capture objects from many angles. The fixed camera location also simplifies the process of determining the location of defects on the piece.

- Another option is to initially create a model that finds easily identified features on the piece. For the power substation example, you could have general instructions on how to point the camera at a transformer in the substation. An object recognition model could identify the bushings on the top of the transformer. This would provide reference points to scale the images with images captured at similar angles. These object recognition models are like facial recognition models that determine the various key points on a face (Long 2018, Long 2019).

- In the case where issues develop or occur over time, a video can create many images, both of known good cases and defect cases.

- You will use the images to create a classification model using convolutional neural networks (CNNs). Depending on how well labeled your data is, you can create models of various complexity.

  o If you have a collection of known good images, you can create a binary classification model that identifies images with a high likelihood of known good or suspected anomaly images. The power transformer is an example of this.

  o If you have images that have been labeled with known defect types, you can create a more complex classification model that identifies the various defects. The discrete parts are an example of this. There might be previous images labeled with an incorrect bearing insertion, and other images labeled with incorrect part milling.

  o If you have good location identification, you can also break down the images and find the locations of portions of the image with defects. The semiconductor wafer is an example here. This would enable you to quantify the expected yield based on the proportion of the wafer with defects.

- When you have trained the model, you can determine at what latency you can test new images as they stream through the model. Determine whether you need to stream image-by-image and get immediate results. Alternatively, you might be able to capture a batch of images and process in batch. Also determine whether the inferencing can be done in the cloud or server or whether an edge gateway is needed.

  o The power transformer example might be a good candidate for edge inferencing. Since the transformers exist in remote locations with reduced or sporadic network bandwidth, an edge device could process the images at the substation. In addition, only rarely is there an issue that requires attention. Processing at the edge would eliminate the need for a large amount of image transfer for rare events.

  o The discrete parts application would be a good example of real-time inferencing in the cloud or server. The factory installation allows high quality network connectivity. The low latency would allow defective parts to be identified immediately and removed before being used in subsequent assemblies.

  o The semiconductor example is a batch process that would fit well with batch inferencing. Since each step in the process is costly, it would be worth the time needed to hold the batch until it could be verified in full detail.

Computer vision is a technique that has many applications for digital twins (Sethi 2018, Gong 2019, Gong 2019). Chapter 11 explores how you can create intelligent computer vision systems, deploy those models on edge-devices to score streaming data, and use SAS Event Stream Processing to make decisions about what is seen in real time.

Applying Recurrent Neural Networks

Recurrent neural networks (RNNs) are a special class of deep learning neural networks designed for sequence or temporal data. An area where RNNs are popular is natural language processing. In this application, large text and document databases are modeled to discover common word sequences in context. The model can then be used to generate new messages for the appropriate context. This is seen in chatbot applications. Within IoT and digital twins, there are many examples of such sequence and temporal data. Many sensors are collecting data over time. The sequence or pattern of the measurements over time can be used to understand interesting characteristics of the digital twin asset. One example is measuring energy circuits in a smart building or power grid. The pattern of the energy use on a circuit can capture the start or end of an asset operation, such as a motor start, which signals an operation change in the digital twin asset. Another use of RNNs is for forecasting unusual time series data. An example is forecasting the energy output from the solar farm referenced earlier in the chapter (Figure 8.12).

Figure 8.12: Solar Farm and Power Output Chart

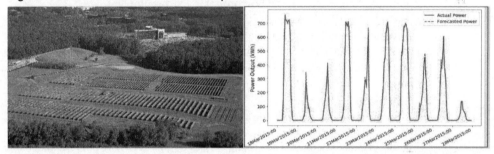

In this case, there is a cyclical component that could be forecasted using traditional methods, but there is a less well-modeled component of weather and cloud cover. With the large amount of data available from the solar farm and nearby solar farms, a deep learning RNN can capture the more sporadic aspects of the energy output (Kahler 2018).

The process for training an RNN is different if you are working with sequence data versus working with temporal data.

The process for training the RNN with sequence data is as follows:

- Break the data into segments of sequential measurements. The length of the segment is determined by the time interval of the data and the expected duration of the precursor to an event. For the energy circuit example in smart buildings, the data is collected at five-second intervals, and we use the previous minute of data.
- Create a target variable for the events of interest and use it to label the sequences where the event occurs. For our example, we are using motor starts and identifying weak motor starts indicating capacitor failure.
- Train the RNN. Note that RNN training has a feature for bidirectional model fitting. That is useful for natural language processing applications where words can be in a different sequence, but still indicate the same event, like a positive product review. Since our measurement data is always moving forward in time, we do not need to use bidirectional model fitting.

- The trained model can then be deployed for inferencing. In most cases, the model inferencing function is sufficiently fast to be used on the real-time measurement stream, either in the cloud, server, or edge device.

The second type of RNN is used to forecast. The example in this case is to forecast the energy output of a solar farm for short time periods in the future (one hour). The key in this case is to create a set of lagged variables for the predictors and the response variable. The response variable is the energy produced. The steps for training this RNN are as follows:

- Take the historical input database and create lagged variables for the predictors and response variable. The number of lags is determined by the time interval of the measurement data and the expected correlation of previous measurements on the forecast time horizon. For the solar farm example, we are producing one-hour-ahead forecasts, and the data over the last few hours is sufficient to capture the primary effects for the forecast. Note that there are a large variety of conditions possible throughout the year and previous observed weather, even though the forecast horizon is fairly short. Since we have a large amount of historical data of the various conditions, the use of an RNN is appropriate for this problem.

- When creating the lags, you should evaluate your data for missing values and consistent time intervals. The ideal case is data collected at consistent intervals with few missing values. If the data has a large number of missing values or inconsistent collection intervals, you can use the TIMEDATA procedure to improve the data for training.

- Since training and evaluating the RNN model is dependent on the sequence, partitioning the data requires more care than typical random partitioning. In this case, we need to preserve the sequence of the data for use in the model creation steps (training, validation, test). The easiest way to do this is to partition the data based on the time variable. Use the earliest historical data for the training data set. Then use the next time partition for the validation data set. Finally, use the most recent data for the testing data set. This works when the performance of the asset has been consistent over the historical data sample. If there have been periods of degraded performance, then it is best to eliminate that data from the data sets used to create the model.

- Train the RNN. Note that some predictors might be estimates that you can capture from other sources. A major factor in solar farm energy forecasting is weather conditions such as cloud cover. In this case, you will want to include these predictors from a source that can provide actual and estimated values.

- For the solar farm example, we are doing short-term forecasting to be used in energy load and generation balancing. Longer term demand planning is done through a separate, more traditional, forecasting process.

- You can use RNNs for one-step-ahead forecasting where the forecast interval matches, or is less than, the desired forecast interval. This yields the most accurate forecast. In some cases, you might need a multistep forecast to forecast future time periods based on the near-term forecast estimates. These forecasts are typically less accurate but can be tested to determine whether they have reasonable accuracy.

RNNs are a valuable deep learning technique for IoT digital twins. (Qi 2018).

Applying Reinforcement Learning Techniques

Reinforcement learning (RL) is a subfield of machine learning and deals with sequential decision making in a stochastic environment. In any RL problem, there is at least one agent and an environment. The agent observes the state of the environment and takes and executes a decision. Environment returns a reward and a new state in response to the action. With the new state, the agent takes and executes another action, environment returns reward and new state, and this procedure continues iteratively. RL algorithms are designed to train an agent through this interaction with the environment, and the goal is maximizing the summation of rewards.

RL has recently got a lot of attention due to its successes in computer games and robotic applications (Hao 2019, Burda 2018). Besides the simple RL applications, there are still few real-world applications of RL to increase efficiency. We studied and did some research to extend an RL algorithm for controlling the heating, ventilation, and air conditioning (HVAC) systems. HVAC includes all the components that are supposed to maintain a certain comfort level in the building.

Buildings consume 30% to 40% of all consumed energy in the world. Any improvement to this rate could result in a huge saving in energy consumption and CO_2 release (US Dept. Of Energy 2008). Advances of the new technologies in recent years have improved the efficiency of most components in the HVAC systems. Nevertheless, still there are several directions to reduce the energy consumption by controlling different decisions on these systems.

There are two general categories of HVAC system: single and multizone. The single zone problem refers to an area that uses an HVAC system (for example, a heater or an AC system that is installed in an office), where the main control decision is the temperature set-point or just the binary action of turning the device on or off. In multizone systems, a central HVAC system supports several zones (for example, offices, hallways, conference rooms), and with a given set-point for each zone, the system needs to maintain a certain comfort level in each zone.

We considered a multizone system and selected the amount of air flow as the main control decision. Using the obtained data from a building at SAS in Cary, NC, we trained an environment and used it to train an RL algorithm when there are 10 zones in the system with a set-point of 72 with ±3 allowance. Figure 8.13 shows the results of 50 cases with different initial temperatures. The upper figure is the temperature, and the lower figure is taken actions more than 150 minutes, in which every three minutes a decision is taken. We compared this result to the commonly used rule-based algorithm (in which the system is turned on/off at 69/75) and RL obtained 47% improvement on combination of obtained comfort and energy consumption.

Figure 8.13: The Average Temperature and Average Action of the RL Algorithm

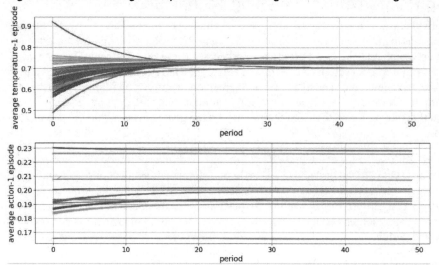

Hyperparameter Tuning

For all deep learning methods, hyperparameter tuning is an important step. Hyperparameter settings are often dependent on the domain knowledge of the application. Research into the specific application can yield a set of parameter settings to be tested. In some cases, a set of parameter settings has been established as best practices. In other cases, research is needed to determine the best settings.

One feature in SAS Visual Data Mining and Machine Learning is hyperparameter autotune. This feature takes a range of potential parameter settings and perform an optimal search for the best performing settings (Koch et al. 2018, Koch et al. 2018).

Conclusion

IoT data coming from physical devices provides the foundation for a digital twin of your remote systems. The data itself provides basic reporting of the current operating state of the device. Adding analytics to this basic digital twin greatly enhances the benefit of capturing this real-time streaming sensor data.

Analytics enable you to fill in gaps that occur due to the limitations of your sensors. Usually, you would like your digital twin to give you information about the overall performance or health of the remote device. Individual sensor measurements can provide some key portions of this overall picture, but analytics capture the relationships and interactions that are significant.

Analytics also alert you when the remote device is not operating normally and enable you to troubleshoot these anomalies. A remote device can enter an undesired condition in several ways. The relationship between parts of the device might be incorrect. An unusual operating mode might be detected, or changes in your remote device can occur over time that can manifest themselves as a different pattern than the normal sequence of events. Various analytic algorithms in your digital twin detect these anomaly conditions.

Use of deep learning is growing into new application areas, and IoT digital twins are an application that will benefit from this growth. The applications for deep learning are different for digital twins than the typical applications seen today like autonomous driving, facial recognition, and natural language processing. Instead, computer vision can be used to perform visual inspection for defect detection and identification. Recurrent neural networks can be used on the time series data streams that are prevalent in IoT to find complex pattern sequences and forecast where effects of environmental variables can be discovered with the large amount of data in IoT. Reinforcement learning can find many applications in IoT as assets in rich environments can be controlled by the algorithm, and results can be measured to determine the benefits achieved.

References

Burda, Y. 2018. "Reinforcement Learning with Prediction-Based Rewards." Available: https://blog.openai.com/reinforcement-learning-with-prediction-based-rewards/

Gong, J. 2019. "Using Deep Learning for Tumor Segmentation in Medical Images." Available: https://blogs.sas.com/content/subconsciousmusings/2019/02/15/using-deep-learning-for-tumor-segmentation-in-medical-images/

Gong, J. 2019. "Constructing the Front of the Computer Vision Pipeline." Available: https://blogs.sas.com/content/subconsciousmusings/2019/02/28/constructing-the-front-of-the-computer-vision-pipeline/

Hao, K. 2019. "The Rise of Reinforcement Learning" in "We Analyzed 16,625 Papers to Figure Out Where AI Is Headed Next." Available: https://www.technologyreview.com/s/612768/we-analyzed-16625-papers-to-figure-out-where-aiis-headed-next/

Kahler, S. 2018. "Using Deep Learning to Forecast Solar Energy." Available: https://blogs.sas.com/content/subconsciousmusings/2018/07/05/deep-learning-forecasts-solar-power/

Koch, P. et al. 2018. "Autotune: A Derivative-Free Optimization Framework for Hyperparameter Tuning." Available: https://kdd.org/kdd2018/accepted-papers/view/autotune-a-derivative-free-optimization-framework-for-hyperparameter-tuning

Koch, P., Wujek, B., and Golovidov, O. 2018. "Managing the Expense of Hyperparameter Autotuning." Proceedings of the SAS Global Forum 2018 Conference. Cary, NC: SAS Institute Inc. Available: https://www.sas.com/content/dam/SAS/support/en/sas-global-forum-proceedings/2018/1941-2018.pdf

Long, X. 2018. "Understanding Object Detection in Deep Learning." Available: https://blogs.sas.com/content/subconsciousmusings/2018/11/19/understanding-object-detection-in-deep-learning/

Long, X., Du, M., and Hu, X. 2019. "Exploring Computer Vision in Deep Learning: Object Detection and Semantic Segmentation." Proceedings of the SAS Global Forum 2019 Conference. Cary, NC: SAS Institute Inc. Available: https://www.sas.com/content/dam/SAS/support/en/sas-global-forum-proceedings/2019/3317-2019.pdf

McGrath, D. 2018. "New Architectures Bringing AI to the Edge." Available: https://www.eetimes.com/document.asp?doc_id=1333920

Qi, Y. 2018. "Recurrent Neural Networks: An Essential Tool for Machine Learning." Available: https://blogs.sas.com/content/subconsciousmusings/2018/06/07/recurrent-neural-networks-an-essential-tool-for-machine-learning/

Sethi, S. 2018. "Computer Vision Using SAS® Deep Learning." Available: https://www.sas.com/offices/pdf/ax-2018-milan/sas-ax18-sethi.pdf

U.S. Department of Energy. 2008. "Energy Efficiency Trends in Residential and Commercial Buildings" Available: https://www1.eere.energy.gov/buildings/publications/pdfs/corporate/bt_stateindustry.pdf

About the Contributor

Brad Klenz is Distinguished IoT Analytics Architect at SAS. His responsibilities include the design and integration of analytics required by SAS IoT solutions. His recent projects include the use of real-time computer vision on edge devices and the use of deep learning for digital twins. He is a patent holder for streaming analytics methods for anomaly detection and event classification in IoT. He is also chair of the Industrial Internet Consortium (IIC) Industrial Artificial Intelligence task group.

Chapter 9: Leveraging ESP to Adapt to Variable Data Quality for Location-Based Use Cases

By Prairie Rose Goodwin

Introduction

The proliferation of smart phones has opened innumerable possibilities for location-based analytics in many consumer industries. Even so, geolocation data reliability can vary wildly with changing environmental factors. SAS Event Stream Processing can detect changes in the quality of your data in real time and adjust accordingly. Its intelligent flexibility keeps up with your customers in real time, enabling you to get the most out of your data.

Why Use Real-time Location Analytics?

The IoT revolution has exploded the number of sensors transmitting information every second. Whereas data used to be static, it is now moving, on the go, and in our pocket, connecting us to the world in new ways that were not possible in the recent past.

As a result, businesses must be able to make sense of the massive amount of data being generated. It's not just big data. It's people's lives and experiences being shared in real time, always growing, always changing. Streaming data holds valuable information about potential opportunities to strengthen the relationship between business and customer, but as the moment passes, and it becomes data at rest, that opportunity is already lost.

Analytics at the edge allows businesses to take advantage of the information available at that particular moment in time. But nothing happens in a vacuum, and context can completely change the meaning of information. Moreover, a business might not get another opportunity to woo a customer if it acts on faulty assumptions. Location adds that context. For example, Figure 9.1 shows baristas and customers in a café. Customers waiting in line for 20 minutes is very different from customers enjoying their beverage for 20 minutes. Location differentiates these two scenarios.

Figure 9.1: Example of Real-time Location Analytics

(Source: https://www.goodfreephotos.com/albums/people/bartenders-and-customers-in-a-coffee-shop.jpg)

Just as analytics at the edge gives businesses a time advantage, geotagged data adds a physical dimension to customer data that can be just as important. The obvious example for location analytics is to use the geotags to separate the data into smaller groups and compare similar information collected in different places. However, the true advantage of geolocation data goes far beyond comparative purposes. It can illuminate cause and effect trends, quantify the efficacy of signage, show inefficiency in staffing assignments, ensure asset recovery, and allow businesses to create customer profiles as they gain insights into their customers' behavior.

But with great power comes great responsibility. Location data is often highly sensitive data because of the personal information that it contains. Businesses must find the right balance between not intruding on the customer's experience while not hiding data collection activities. Big Brother is real, and it is technology that passively records every move. (In this context, passive implies that the infrastructure does not require the user's intervention to collect data.) Most people know that their data is being collected, but very few appreciate just how much personal information can be gleaned from it. Therefore, legal, ethical, and practical implications should be discussed before deciding that location tracking is right for your business.

Location and Privacy

Before collecting customer data, every business should fully understand the legal, ethical, and business implications of what they are doing. There is no question that smart phones have opened a Pandora's box of peeking into people's personal lives. Everything from their frequently visited establishments, to their browsing history, to when they go to bed every night can be tracked by the behemoth companies that control the smartphone market. This data is sold to advertisers to be used for targeted ads. However, there is beginning to be more widespread concern about privacy as it becomes more well-known just how much data is generated from every move (or even lack thereof).

The European Union passed a sweeping set of data privacy regulations known as the General Data Protection Regulation (GDPR). The law requires informed consent from the person before data is collected; otherwise, it must be completely anonymized. The problem with that is that it is very hard to make location data completely anonymous if it spans multiple addresses. Knowing where someone sleeps most nights tells you someone's address, and a quick Google search can even return the name of the person who lives there. If you see the device travels regularly to a middle school via a bus route, there is a good chance that device belongs to a minor, which brings up new issues in how businesses should interact with that device. When dealing with sensitive information that includes someone's whereabouts, businesses must be careful that what they are doing is both legal *and* ethical based on what information can be gleaned about that person.

Indoor location tracking is less invasive because it narrows activities into one specific context in which that person is interacting with the business. However, you still need to worry about the Big Brother effect. Customers generally do not like feeling like they are being watched and might avoid the establishment all together or change their behavior in order to hide their intentions from the establishment.

Types of Location Data

There are three main types of location data: outdoor geolocation data, categorical data, and indoor location tracking. *Outdoor geolocation data* takes the form of an address, area, or longitude and latitude. These data points are useful in large scale use cases such comparing the pollution in a city with the amount of childhood asthma to find any correlations. Data sets in this category can cross-reference trends with streets, city boundaries, points of interest, demographics, income, transportation routes, schools, and cost of living to name some examples. Organizations look at all these factors to inform their decisions so that they serve the needs of their communities effectively. By placing a storefront in an optimal location, pricing their products within reach of the average income, they are more likely to establish a lasting presence in that area.

The second type of data acts as *categorical information* and does not necessarily lend itself to being mapped. For example, at a conference, if each booth has a badge scanner, the resulting data set would have a list of places that attendee had gone, but the data would not contain a way to link the datapoints together accurately. This type of data is generally text-based, but it can be mapped to X, Y locations if and only if there is an external mapping available. Conversely, if the data must be anonymized, it can be put in this form to remove identifiable locations, which could jeopardize the purpose of anonymizing the data in the first place.

The third type of location data is *indoor location tracking* that results in x,y coordinates in a custom mapped space. Whereas GPS gives you a particular point in space that is the same across all GPS data, indoor tracking is only relevant to the environment in which it was collected. This type of data is most well fitted to small-scale use cases that are specific to an event or place. It also requires special hardware infrastructure. GPS tends to break down in buildings due to the interference of the structure between the device and the satellite. Whereas outdoor tracking can be accurate to a few meters, the quality of indoor location tracking is unreliable at best and non-existent at worst. Because of this, indoor tracking is the most difficult category of data to collect accurately, but it is also the most important for businesses.

This chapter focuses on the third type of data, indoor location tracking, although the principles discussed are also applicable to the other two (Figure 9.2).

Figure 9.2: Three types of location data collection. a) GPS data: usually latitude and longitude, best used for outside use-cases; b) RFID chip: records when present at a scanner; and c) indoor tracking: x,y points, often done through WiFi access points

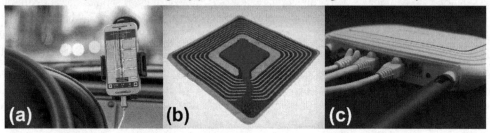

(Source: (a) https://www.pexels.com/photo/smartphone-car-technology-phone-33488/, (b) https://upload.wikimedia.org/wikipedia/commons/4/40/RFID_Chip_004.JPG, (c) https://pixabay.com/photos/wireless-home-router-adsl-modem-1861612/.)

Location Collection Technologies

There are two different ways to collect location: active and passive. *Active tracking* requires the user to self-report their location. For example, some buildings require individualized key codes or badges that unlock doors. Alternatively, businesses entice some customers by offering rewards for checking in. In both cases, the act of tracking is front and center, which can cause customers to modify their behavior, and it is not very accurate because people are more likely to opt out. Active tracking is generally too difficult to implement reliably for any but the most secure locations.

Passive tracking is much more valuable because it elicits natural behaviors and does not put customers on guard. Almost all businesses could benefit from the insights gained from passive tracking, but collecting it is not necessarily an easy or straight forward task. It is possible to do it with individualized tags such as RFID tags, but smartphones provide a valuable resource that can be leveraged automatically. However, even within the smartphone space, there are multiple technologies that can collect location passively.

The most common approaches to passive location tracking are discussed below. This list is not meant to be exhaustive, and there are variations within each individual technology.

GPS

GPS (Global Positioning System) is the most well-known way to collect location data. This technology works by receiving signals from one or more satellites orbiting the earth. Most smart phones are GPS-enabled, and location services allow third parties to tap into the geolocation data easily. The accuracy of GPS in the outdoors can vary significantly, but it is often good enough for most outdoor use cases – usually within a few meters, and almost always within building-sized areas. However, structures interfere with GPS signals, making it less reliable indoors. For those use cases, a different technology should be used.

WIFI

WiFi location tracking is a good candidate for venues that already offer wireless internet access to their customers. The types of environments that would need location tracking are likely to have multiple internet access points distributed throughout the space. When a device connects to the network, specially equipped access points within range measure the Received Signaled Strength Indication (RSSI). The position of the device can then be triangulated within a few meters, or approximately accurate to within a room-sized area. However, care should be taken when deciding where to place the access points to complement the requirements of the environment. Interference is still possible, and an incorrectly installed WiFi-based tracking system will not provide reliable data.

Bluetooth Beacon

Bluetooth beacons are becoming more commonplace for advertising use cases. For example, these low-energy devices are distributed throughout a store and deliver targeted coupons related to the products around the beacon. Similar to WiFi tracking, location can be determined by triangulating a receiver's signal by measuring the RSSI at multiple beacons. Accuracy using Bluetooth is also comparable to WiFi, as is the issue of environmental interference.

Passive RFID

Passive RFID is the most accurate collection method discussed because it has the shortest range of signal, between one and five meters, usually within a table-sized area. However, unlike the other technologies discussed, it requires special hardware not present on a smartphone: the tag that moves and the reader that is usually stationary. The readers can report seeing the tag within its range, but unlike the other technologies, the data can be translated to X, Y coordinates only if you map it to the X, Y location of the reader. Often used for theft prevention, RFID is low cost, but is the most visible methodology discussed in this chapter, which makes it the least passive.

Use Cases

Location is valuable in many different contexts, but this chapter focuses on how businesses can leverage this information while interfacing with customers. While this probably evokes ideas of retail spaces, that is far from the only scenario. Conferences are incredibly important for building relationships and offer a plethora of metrics by which to judge the efficacy of the event. Corporate transactions happen not in a store, but in the facilities that house the same employees that keep the business running every day. In all three use cases, it is important that everything is running smoothly to fully optimize the customer experience.

The following use cases are customized for business and customer interactions but can be useful to other environments as applicable.

Asset Recovery

A significant amount of resources goes into providing and maintaining hardware in the care of employees. When one of those assets goes missing, it can mean days of productivity lost for that worker in addition to the cost of replacing the item. By having assets connected to a location tracking system, they can easily be pinpointed, assuming the assets are still connected. If not, the last recorded location shortens the search significantly. If that person is only a visitor, as in the case of a conference, it could be the difference between getting the device back and having to replace it.

Customer Engagement

Customer engagement can be defined as customer interaction with the environment. Retail spaces depend on customers to browse merchandise, read a sign, or interact with employees. Products must be strategically placed so that customers continue to look at other items in addition to the one that brought them to the store. Similarly, a sign that no one sees is not worth the paper that it is printed on. Figure 9.3 is an example of a customer engaging with a window display in retail area. Knowing what windows receive the most foot traffic can help products be discovered at a glance.

Figure 9.3: Example of Customer Engagement

(Source: https://www.pexels.com/photo/woman-wearing-gray-coat-935760/)

For conferences, customer engagement takes the form of people visiting booths. Booths make up an important aspect most business conferences – one that represents a significant amount of resources invested in terms of space, resources, transportation, setup time, and breakdown time. Booths that do not spark any engagement are a drag on the entire event. That lack of engagement can ripple to nearby booths and suck the energy out of the room.

All important customer engagements take time to happen. Therefore, it can be seen with high-quality location tracking. Dwell time is a good indicator of whether customers are engaging with something in that particular area, whether it be a product or a booth. Similarly, the process can be reverse engineered such that if you want people to engage with a particular thing, you can move it to a location with lots of foot traffic to improve visibility.

Staffing Priorities

Staffing can be difficult to judge beforehand. Areas that you might expect to be popular might not be getting the predicted amount of attention while other parts might be slammed beyond what the staff can handle. Being able to track where customers spend their time can enable you to more effectively service the needs of your customers.

Staffing priorities can be interpolated by looking at "hot" zones and comparing how the percentage of customers in each zone compares to the percentage of staff in each zone. If the two heat maps do not look the same, it might be worth it to move staff around to better match where the attendees are spending their time. Moreover, tracking the number of people over time will tell you what the peak hours are for your event. This information can inform when to give staff breaks versus when more people might be needed.

Space Utilization

Space can be the most expensive asset to consider, especially when working in expensive urban areas like New York and San Francisco. In these cases, maximizing every square foot is paramount. Similar to staffing priorities, the reality of how a space functions might not exactly match what you expected. Being able to see how customers move through a space is incredibly valuable and can be used to analyze your current space usage and give you solid metrics by which to inform decisions. For example, Figure 9.4 shows professionals networking at a conference. Knowing what areas are utilized at conferences can help events make the most of their available space.

Figure 9.4: Example of Space Utilization

(Source: https://www.dreamstime.com/people-inside-cafe-tables-chairs-public-domain-image-free-109928269)

Cause and Effect

At conferences, there are usually many events that attendees can choose from. Though it is useful and straightforward to know which events are popular, it is even more useful to be able to track cause and effect from other parts of the conference. For example, if 75% of people who visit a booth then attend a talk by that person as opposed to 25% for other speakers, the relative popularity of that event is high even if they do not have the highest raw attendance numbers. Another important connection could be behavior of attendees who decided to follow up with sales. By tracking who they talked to, you could begin to understand which staff members are the most effective sales representatives.

This type of analysis is largely done offline after the event is over, instead of purely in real time, because you have the benefit of being able to look both backward and forwards and do multiple passes on the data to fully use the information at your disposal. Linking events can be as simple as taking a metric, such as those attendees that became customers, and seeing what if any similarities exist in their conference experience.

Customer Profiles

Customer profiles are the most data-intensive use case for location tracking because they require knowing context about the environment, the customer, or preferably both. Profiles go beyond explaining *what* the customer is doing and begin to answer the question *why* they are doing it. For example, if your event has a student session and a job fair session, and you see someone attend both, you might begin to assume that that person is about to graduate. On the other hand, if you see someone attend a student session and an executive management session, that person is more likely to be on the other side of the table looking to hire someone.

Data Variability

Incomplete or too sparse data can unravel even the best statistical methods. Conclusions that are rock solid in some circumstances become unusable in others. For location tracking, steady data streams can disappear suddenly if the device in question goes into a power-save mode or leaves the trackable area. Missing data is not just commonplace, it is inherent in the way networks currently communicate and needs to be accounted for.

Potential Problems in Data Collection

The model discussed in this chapter focuses on passive collection methods that resolve to X, Y coordinates but can vary significantly over the course of collection. Some errors can be accounted for during analyses, but some are directly related to the quality of the installation and how it matches the unique considerations in a space. If the setup is not ideal, the data can still be used, but it should be accounted for with a confidence interval. Here are some issues to be aware of:

Installation Issues:

- **Access points that are too far apart.** Earlier, we described two different approaches that use the RSSI to triangulate the device's location between multiple access points. This requires access points to be placed evenly apart to get consistent coverage throughout the venue.
- **Large windows.** If your space contains large windows, research how that affects the type of signal produced by your location tracking systems. Some types of signals can be reflected by glass showing interference or doubles in your data set.
- **Signals from other floors.** WiFi in particular is omni-directional. If access points are placed in the ceiling in a multifloor building, they will likely pick up devices above them as well as below.

Data Set Issues:

- **MacAddress scrambling to protect the owner's privacy**. Privacy is becoming more and more of a concern in our ever-connected world. Smart phone manufacturers have responded to this problem by hiding the device's real MacAddress from untrusted access points, changing it every minute. It can be difficult if not impossible to link datapoints together that come from the same device. Only some analyses can be run on this data. Therefore, when possible, making people authenticate to the network instead of only logging in as a guest improves your data significantly. (Internal experiments showed approximately 70% of our collected data came from devices that scrambled their MacAddress.)

- **Walking by an area does not indicate engagement.** If there are many sensors communicating information, this is important to control for. In these cases, the data contains a lot of devices that record only a couple of pings (depending on how often you are receiving information). The dwell times of these instances is about equal to the data collection interview.

- **Devices can go into power saving mode**. Unfortunately, system administrators do not have as much control over the data collection process when passively tracking from a smartphone. Some devices that can produce great quality and density might suddenly stop doing so. This is because phones go into a power Save mode when not in use.

- **Ghosts.** Passive communication is largely initiated by the smart phone, which means it can be difficult to ascertain if the device is still in the space but in power Save mode, or if it has actually left. Decide what threshold is appropriate for your particular use cases to remove devices after they have stopped communicating.

Leveraging SAS Event Stream Processing to Adapt

Data density and count are the two most important facets to measuring your data quality. SAS Event Stream Processing can detect both in real time, which can change your confidence intervals. For example, about 70% of smartphones currently use MAC address scrambling, which makes it impossible to figure out which datapoints came from the same device. This immediately invalidates 70% of your data for path planning. ESP can detect this issue, and instead of throwing all the data out or using it improperly, it can filter it into customized analyses, while still giving you real-time path planning for the remaining 30% of data.

Model to Passively Track Staff and Customers

In this example, we are looking at a cafeteria that tracks staff and customers passively with a WiFi-based system. The system administrator wants to know what the person is doing, a rough estimate of the number of people in the building at the time, whether there is a bottleneck at the registers that requires more staff, peak hours, which stations are the most popular, and label any data with user information for other uses. All of this can be done in one pass in real time.

Model to Analyze Data Quality

To do everything our system administrator wants, SAS Event Stream Processing needs to filter out the data that does not match his or her goals. That data not used in this model can be easily saved offline at any step for later use. Figure 9.5 breaks down each step of the model individually by goal.

Figure 9.5: A Visualization of Data Flow Through the Example ESP Model for Location Tracking

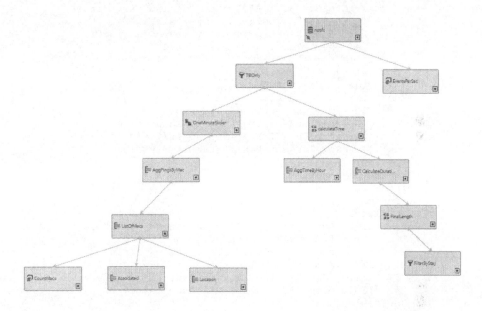

The individual steps of the model are as follows:

1. **Reduce geolocation to specific area.** Some systems give you the zone based on which access point was used to collect the data. Otherwise, it can be done with geofencing the raw X, Y coordinates. If your data is tagged with geofencing, this can be done in a single line in a filter window.

    Example Code 9.1

    ```
    <window-filter name="TBOnly">

    <expression><!match_string(locMapHierarchy,'*TB*')></expression
    >
    </window-filter>
    ```

2. **Calculate duration of stay.** To calculate the duration of stay, take the timestamp of the last datapoint associated with a MacAddress and subtract it from the first. This continues to update with more incoming data points using an aggregate window.

Example Code 9.2

```
<window-aggregate index="pi_HASH" pubsub="true"
name="CalculateDuration">
  <schema>
    <fields>
      <field name="deviceID" type="string" key="true"/>
      <field name="associated" type="string"/>
      <field name="locMapHierarchy" type="string"/>
      <field name="arrival" type="int64"/>
      <field name="lastSeen" type="int64"/>
    </fields>
  </schema>
  <output>
    <field-expr>associated</field-expr>
    <field-expr>locMapHierarchy</field-expr>
    <field-expr>ESP_aFirst(timestamp)</field-expr>
    <field-expr>ESP_aLast(timestamp)</field-expr>
  </output>
</window-aggregate>
<window-functional name="FinalLength">
  <schema>
    <fields>
      <field name="deviceID" type="string" key="true"/>
      <field name="associated" type="string"/>
      <field name="locMapHierarchy" type="string"/>
      <field name="arrival" type="int64"/>
      <field name="lastSeen" type="int64"/>
      <field name="StayDuration" type="int64"/>
    </fields>
  </schema>
  <function-context>
    <functions>
      <function name="StayDuration">
      quotient(diff($lastSeen,$arrival),60000)
    </function>
    </functions>
  </function-context>
</window-functional>
<window-filter name="FilterByStay" index="pi_HASH">
  <expression><!StayDuration > 5 AND StayDuration <
180></expression>
</window-filter>
```

This step has multiple uses. First, we know that staff is there the whole time that the cafeteria is open, which would be abnormal behavior for a customer. Therefore, we can do some basic segmentation. To filter out the staff, we can reduce our data set by only looking at devices that were on-premises for less than three hours.

Secondly, for this use case, we do not care about pass-through people. Therefore, we can set the minimum duration of stay at five minutes. This also removes all scrambled MacAddresses, which will show up as a single datapoint.

At the end of this step, the model has a collection of MacAddresses that meet the criteria of staying between five minutes and three hours in the cafeteria. This window needs to be merged with the last step in order to get all the location datapoints associated with those MacAddresses.

3. **Calculate peak hours.** The next branch of the model calculates the peak hours by calculating the hours the person was there from the raw timestamp, and then aggregating the results by the hour. This can be easily visualized with a bar chart.

Example Code 9.3

```
<window-aggregate index="pi_HASH" pubsub="true"
name="AggTimeByHour">
  <schema>
    <fields>
      <field name="hourOfDay" type="double" key="true"/>
      <field name="Counter" type="int32"/>
    </fields>
  </schema>
  <output>
    <field-expr>ESP_aCount()</field-expr>
  </output>
</window-aggregate>
```

4. **Remove outdated information.** Knowing when someone leaves a space can be difficult in passive tracking. Smartphones ping at regular intervals when in use, but most devices go into a power-save mode during inactivity, which can greatly increase the intervals between pings. So how can we tell what devices have gone to sleep and which ones have left the building? You cannot. However, in SAS Event Stream Processing, you can set a threshold at which point it assumes that person has left. In this model, we have used a 30-minute threshold by including a sliding window, at which point devices are no longer considered active.

Example Code 9.4

```
<window-copy pubsub="true" name="thirtyMinuteSliding">
  <retention type="bytime_sliding"><!30 minutes></retention>
</window-copy>
```

5. **Segment data by behavior for customer profiles.** All the steps up to this point have been either big picture or filtering the data that meets very specific criteria. Once the data reaches this point in the model, we can begin doing more in-depth analyses. The system administrator wants to take a best guess at what customers are doing. This can be accomplished by setting up geofenced zones that indicate particular stages of lunch: standing in line to get food, checking out at the registers, eating in the dining area, and so on. By having an enumerated variable that tracks the current stage and rules for transitioning to any other geofenced location, SAS Event Stream Processing can give the system administrator its best guess for activities going on at any given time.

Example Code 9.5

```
<window-functional pubsub="true" name="BehaviorSegmentation">
  <schema>
    <fields>
```

```
            <field name="deviceID" type="string" key="true"/>
            <field name="timestamp" type="int64" key="true"/>
            <field name="x" type="double"/>
            <field name="y" type="double"/>
            <field name="confidenceFactor" type="double"/>
            <field name="locMapHierarchy" type="string"/>
            <field name="associated" type="string"/>
            <field name="stayDuration" type="int32"/>
            <field name="lastSeen" type="int64"/>
            <field name="segment" type="int32"/>
            <field name="togo" type="int32"/>
        </fields>
    </schema>
    <function-context>
        <functions>
            <function name="segment">0</function>
            <function
name="togo">if(lt(stayDuration,20),1,0)</function>
        </functions>
    </function-context>
</window-functional>
```

6. **Calculate current occupancy.** As mentioned in the previous section, calculating occupancy with passive tracking is possible, but it should be used cautiously. Scrambled MacAddresses appear as different devices every minute. Therefore, if you have not already filtered out those datapoints, you can look only at one minute of data for this use case. Alternatively, if you have already filtered out this data, the results are a lower bound. Another approach would be to use SAS Event Stream Processing to calculate the percentage of scrambled to non-scrambled MacAddresses and give you an estimate of the total occupancy by combining that information with non-scrambled MacAddresses. In all cases, this number is an estimate only, best used for comparison purposes.

Example Code 9.6

```
<window-copy index="pi_HASH" name="OneMinuteSlider">
  <retention type="bytime_sliding"><!60 seconds></retention>
</window-copy>
<window-aggregate index="pi_HASH" pubsub="true"
name="Location">
  <schema>
    <fields>
      <field name="MacAddress" type="string" key="true"/>
      <field name="Counter" type="int32"/>
    </fields>
  </schema>
```

7. **Identify "hot" areas.** Once historical data has been collected, you can begin looking for anomalies that indicate problems. For example, if only 50% of the cash registers are open and that area begins to get overcrowded, SAS Event Stream Processing can send an alert to the system administrator, so that he or she can send in more staff. If no staff is available, the administrator could look to see whether there are cold spots where someone can be moved while minimizing the effect of their absence. Alternatively, marketing can be placed strategically so that it has the most exposure

possible. Knowing how your space is used is key to being able to know how your space can be optimized.

Example Code 9.7

```
<window-copy index="pi_HASH" name="OneMinuteSlider">
  <retention type="bytime_sliding"><!60 seconds></retention>
</window-copy>
<window-aggregate index="pi_HASH" pubsub="true"
name="Location">
  <schema>
    <fields>
      <field name="locMapHierarchy" type="string" key="true"/>
      <field name="Counter" type="int32"/>      '
    </fields>
  </schema>
  <output>
    <field-expr>ESP_aCount()</field-expr>
  </output>
</window-aggregate>
```

8. **Associate more information.** To fully capitalize on all the advantages of location tracking, the data must be saved and associated with additional information such as customer IDs, purchase history, previous location data, job title or description, and so on. If using WiFi collection, being connected can allow the system to collect this information with a sign-in page. Stores like Target give customers incentives to download their proprietary apps, which open up even more avenues for passive data collection. At this point, it would be saved to a backup server where it could be leveraged at a later time.

Conclusion

The importance of analytics cannot be understated in the modern marketplace. Big data has opened up a new world of connecting with customers and to not leverage that information is to lose that business to a competitor. In an ever-connected world, decisions need to be made immediately and with as much information as possible.

Location is changing how we look at data by providing important insights. As a business, you need to understand what factors are affecting your business and how you can use them to improve your bottom line. To do this, organizations have to decide what infrastructure best fits their needs. They need to determine what use cases on which to focus.

We have seen SAS Event Stream Processing handle use cases that have few data requirements to use cases that only work if the data meets a very specific set of criteria. As data traverses the model, it is leveraged in different ways to get the most out of the data. As the data changes, it is rerouted so that it fits the use case, rather than losing out on valuable real-time insights. SAS Event Stream Processing provides real-time analytics, which leverages the power of time to put your business one step ahead.

About the Contributor

As a Senior Product Developer in the IoT Division at SAS, **Prairie Rose Goodwin's** work focuses on using the power of real-time analytics on the edge to deliver important insights when it matters. She holds a BA from Vassar College and completed her MS and PhD at North Carolina State University.

Chapter 10: Condition Monitoring Using SAS Event Stream Processing

By Anya McGuirk, Yuwei Liao, Byron Biggs, Deovrat Kakde, and Joseph Costin

Introduction

Internet of Things (IoT) technology enables *condition monitoring* of critical, high-valued machines. Measurement and monitoring of one or more health parameters is an essential component of condition monitoring. It is now generally accepted that condition monitoring, where possible, is economically superior to the practices of *preventative maintenance* – conducting maintenance in predetermined, fixed-time intervals. It is also superior to the practice of *run-to-failure or corrective maintenance*, which refers to running the machine until it fails. But understanding the internal condition or health of a machine is not trivial, especially when the machine is constantly in operation, as is often the case.

One popular way to monitor machine health is by analyzing vibration data. *Vibration analysis* requires a vibration sensor on the machine that can take multiple readings per second. The real-time analysis of such data requires analytical tools that can handle a large volume of data at sub-second rates and turn the data into actionable insights in seconds.

This chapter explains how we investigated the following research questions:

- Can our more advanced analytical methods handle the large data volume typical of streaming sensor data?
- What SAS tools are best to detect faults using vibration data?
- Which is more effective: time or frequency domain analysis?
- If the motor runs at different speeds, do we need different models?
- Can we detect faults and use these tools in real time with data streaming at very high sampling rates?

We show how we use SAS Event Stream Processing to monitor machine health with data streaming online at high sampling rates.

Experimental Setup

To conduct the research, we used a variable speed, three-phase Induction Squirrel Cage Motor equipped with three accelerometer vibration sensors (Figure 10.1). Using this machine, we ran three experiments with the motor running at 25, 35, and 50 revolutions per minute (RPM), respectively.

Figure 10.1: Variable Speed Squirrel Cage Motor

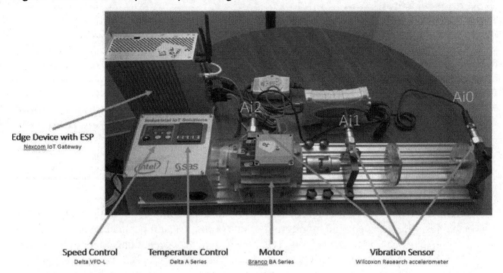

While running each experiment, we collected data from the vibration sensors at 12,800 Hz (that is, 12,800 data reads per second). Initially we collected approximately ten minutes of data with the machine running under its presumed non-fault "normal" state. After approximately ten minutes, we artificially changed the state of the motor from this normal condition. To do this, we adjusted the black knob on top of the blue plate near sensor Ai1 (Figure 10.1). This adjustment effectively changes the alignment of the main motor shaft ever so slightly from its "balanced" state.

To understand what happens as this knob was adjusted, see Figure 10.2. A twist of the knob raises the center rectangular section up slightly, changing the alignment of the shaft that runs through the hole illustrated in the center of this piece. In this way, we caused the shaft to become unbalanced. We then let the machine run in this new state with an artificially induced misalignment.

Figure 10.2: Knob to Adjust Shaft Balance

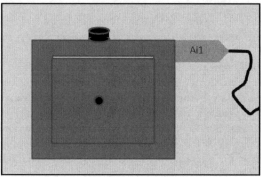

Time Domain Analysis of Vibration Data

We first focused on data from when the motor ran at 25 RPM. We also focused on data from the Ai1 sensor because it is located on the blue plate where we intervened and distorted the balance of the motor shaft. Figure 10.3 shows the time-domain waveform from the "normal" state (light) followed by the data obtained after we adjusted the state of the motor (dark). Time (in minutes) is measured on the horizontal axis, and displacement due to vibration, referred to as the amplitude, is measured on the vertical axis.

Figure 10.3: Raw Sensor Data from Ai1 Over Time

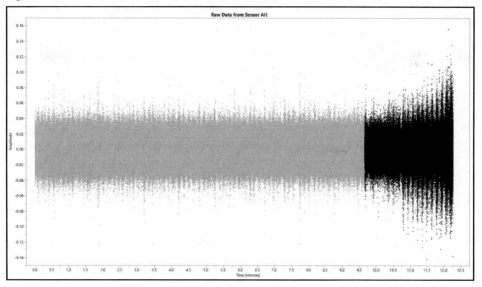

Figure 10.3 shows that, in the normal state, the amplitude of the time waveform is relatively stable. After the new state is introduced (dark), the pattern slowly changes. By minute 11, the change of state appears obvious to the eye.

We started our vibration data analysis by first focusing on the signal waveform, performing analysis in the time domain. When monitoring vibration, the root mean square amplitude (RMSA) of the signal is often tracked. If we let x_i denote the ith vibration sensor measurement, then RMSA is defined as follows:

$RMSA = \frac{1}{N} \sqrt{\sum_{i=1}^{N} x_i^2}$, where N is the total number of measurements.

The data in Figure 10.3 suggests that in addition to tracking RMSA, we might also want to track the kurtosis of the vibration data. *Kurtosis*, the fourth moment of a distribution, measures the thickness or thinness of the distribution tails (the proportion of the observations away from the mean of the series). In Figure 10.3, for example, we can observe the tails getting fatter the longer the machine runs in the altered state. Here, we use the standard formula to calculate kurtosis.

To track both the kurtosis and RMSA over time, we calculated both statistics within a one-second window of the signal ($N = 12800$ observations), and then recalculated their values by sliding the window half a second (window overlap ratio = 0.5). The windowed estimates of RMSA and kurtosis over time are shown in Figure 10.4. To determine where an alert might be thrown, we used the first 75% of the "normal" data to estimate the standard deviation of each series. We then drew the +/- 3-sigma boundaries on the respective graphs. (See the dark lines in Figure 10.4). Based on the 3-sigma thresholds, an alert would likely be thrown at 10.78 minutes, as illustrated by the vertical black line in Figure 10.4.

Figure 10.4: Root Mean Square Amplitude and Kurtosis over Time with 3-Sigma Bounds

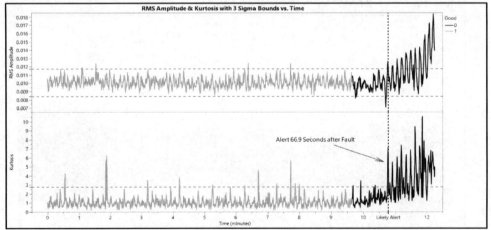

By overlaying the time domain alert on the original data (Figure 10.5), you can see that the alert happens just before our eyes would have noticed a problem.

Figure 10.5: Time Domain Alert

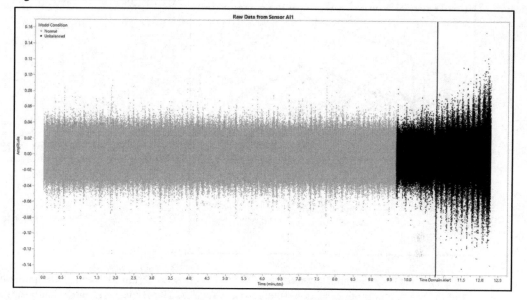

Monitoring Specific Frequencies Using Digital Filters

Some signals are better understood in the frequency rather than the time domain. Vibration signals from rotating equipment are a good example. Rhythmicity or cycles arise from components or parts that rotate. An in-depth discussion of the relationship between the time and frequency domain is beyond the scope of this chapter. Nevertheless, Figure 10.6 provides the intuition that you need to understand the analysis that follows.

Thus far, we have focused on analyzing the amplitude of the waveform in time-domain as seen in Figure 10.6. As illustrated in this same figure, the time waveform can be decomposed into the frequencies that it consists of in order to obtain its spectrum.

The (discrete) Fourier spectrum describes the magnitude of the various frequencies comprising the signal. The frequencies that you would expect to see in the data depend on characteristics of the motor being monitored, such as the following:

- the speed of rotation or RPM
- the diameter of the rotating components such as the shaft
- the size and number of ball bearings

For example, to monitor the state of the inner-race that the ball bearings of the squirrel cage motor ride on, you should monitor frequencies around 53 Hz (assuming that the motor is running at 25 RPM). We arrive at this frequency because the machine has seven ball-bearings that are 12mm in diameter, with a pitch diameter of 32mm, and a contact angle of 15 degrees. Similarly, you can use the motor design and specifications to derive which frequencies need to be monitored to verify that other components are working properly. Once the frequencies of interest are identified, they can be monitored using a digital filter.

Figure 10.6: The Relationship Between the Time and Frequency Domains

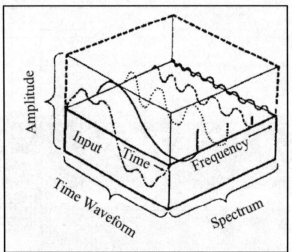

A digital filter takes a signal (input data), selectively chooses certain frequency components from the input, and produces a new filtered output series. The Butterworth and Chebyshev Type I and II digital filters are available in SAS. For our shaft imbalance experiment, we used a high pass Butterworth filter. The high pass filter that we implemented is designed to capture the frequencies 3840 Hz and higher from our original raw sensor data (x_t). The new filtered data is denoted y_t.

Why did we choose this particular frequency band for our experiment? Because we did not know the exact consequences of our intervention on the state of the machine, we first analyzed the raw sensor data (in Figure 10.3) using Short-Time Fourier Transform (STFT) available in the Time Frequency Analysis (TFA) package in the SAS Visual Forecasting product on SAS Viya.

STFT takes a window of data (hence "short-time") and computes the discrete Fourier transform of the windowed signal. That tells us what frequencies are most dominant in the data in that window of time. Then the window slides in time, and the Fourier spectrum of the next window of data is computed.

For our analysis, we used a window size of 4096 observations and then advanced 2048 observations before computing the next STFT. The results of the STFT are shown in Figure 10.7.

Figure 10.7: Short-time Fourier Transform on Raw Sensor Data (Ai1)

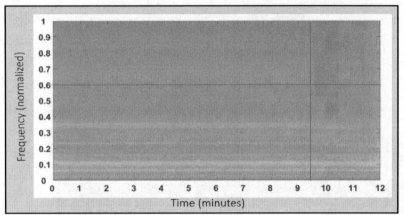

The shading in the figure indicates the power/magnitude of the various frequencies (vertical axis) versus time (horizontal axis). The brighter the color, the higher the power/magnitude. The vertical line in Figure 10.7 indicates where we intervened to change the state of the motor. The horizontal line indicates the cutoff frequency for our high pass filter (3840 Hz or 0.6 normalized frequency). A significant change in the magnitude of the higher frequencies after our intervention is clear in Figure 10.7. We chose our filter cutoff frequency based on this observation.

The output series from the digital filter (y_t, in light color) and the original sensor data (x_t, in darker color) are shown in Figure 10.8. The stability of both the original sensor data and the filtered series before the intervention are evident. The change in model state following our intervention is evident almost immediately.

Figure 10.8: Raw Sensor Data (Dark) and Filtered Series (Light)

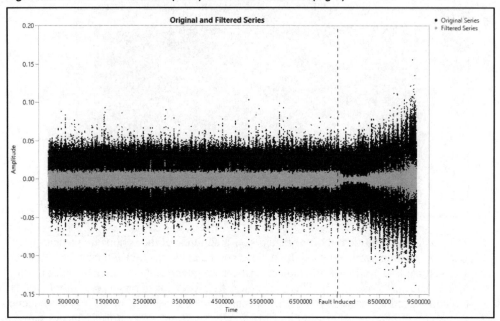

To compare the change in the filtered (high frequency) series relative to the original series, we calculate the ratio (R) of their relative root mean square amplitude as follows:

$$R = \frac{\frac{1}{N}\sqrt{\sum_{i=1}^{N} y_i^2}}{\frac{1}{N}\sqrt{\sum_{i=1}^{N} x_i^2}}$$ using windows of size $N= 512$ observations.

Figure 10.9 shows the full ratio series, and Figure 10.10 depicts a zoomed view of the series where we show where a likely fault would have been thrown. Figure 10.10 illustrates how the alert based on the ratio RMS amplitude is much earlier than the alert based solely on the time domain analysis (kurtosis and RMSA). Both the STFT results (Figure 10.7) and the results based on the filtered series (Figure 10.8 and Figure 10.10) indicate that considering the frequency domain can be very fruitful, especially when you know what frequency range should be monitored.

Figure 10.9: Ratio of RMS Amplitude: Filtered to Original Series

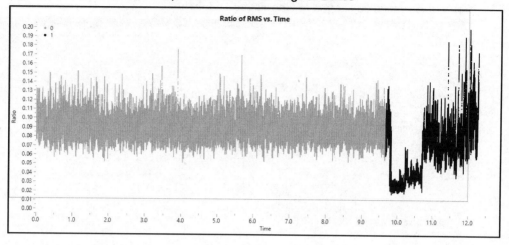

Figure 10.10: Likely Alert Based on Ratio of RMS Amplitude

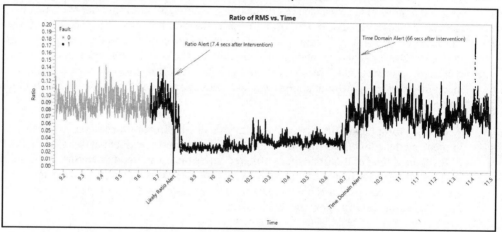

Real-time Analysis Using SAS Event Stream Processing

Analysis of the window estimates of both RMS amplitude and kurtosis is possible in SAS Event Stream Processing even when data is streaming at very high frequency such as 12,800 Hz. Display 10.1 (left panel) shows a screen capture of a working Streamviewer window. Streamviewer is a graphical user interface that allows the user to visualize streaming events.

Display 10.1: SAS ESP Streamviewer: Streaming RMS Amplitude and Kurtosis (Left), Ratio RMS Amplitude Filtered to Non-Filtered (Right)

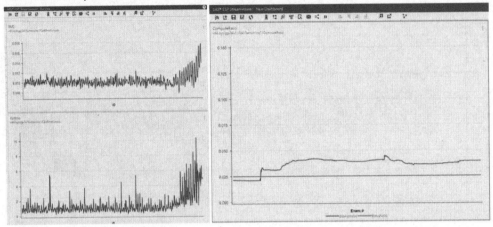

You can easily add threshold values and program SAS Event Stream Processing to throw an alert if the threshold is exceeded. Note that the Streamviewer window statistics are not calculated using overlapping windows. Consider the case here where we use a window size of 12800 observations. In this case, 12800 observations are collected as the data are streaming in, the statistic is calculated, and the data discarded. The next 12800 observations are then collected, and the statistics recalculated. As Display 10.1 (left panel) shows, the graph is not much different from the overlapping window statistics calculated in SAS Viya (see Figure 10.4).

The Butterworth digital filter is available in SAS Event Stream Processing, so we also implemented the high pass filter described above. We then calculated and displayed the ratio of RMS amplitude of the filtered to unfiltered series on streaming data in Streamviewer. (See the right panel of Display 10.1.) In SAS Event Stream Processing, it is easy to trigger an alert after the ratio exceeds a specified threshold. In both cases, as demonstrated in Display 10.1, the high data streaming rate is easily handled.

Monitoring the Whole Fourier Spectrum

Instead of, or in addition to, monitoring a frequency band, you can monitor the whole Fourier spectrum to look for *any* changes in the spectrum. To do this, observe the motor operating in a normal or fault-free state, compute a "normal" or reference Fourier spectrum, and then compute the spectrum for incoming data and look for differences from the reference.

The process of determining the reference spectrum is outlined in Figure 10.11. The reference spectrum is computed offline by analyzing vibration sensor data under "normal" conditions.

Figure 10.11: Estimating a Reference Spectrum

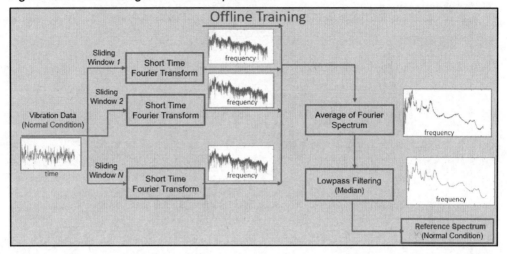

For our experiments, we used the first minute of data under normal operation at each speed to derive a reference spectrum. For the 25 RPM data, we calculated the STFTs as described earlier (windows of length 4096, with overlap 2048). We then averaged the STFTs computed over an interval of one minute (375 windows) to get the average Fourier spectrum. We then smoothed this spectrum using a median filter. The result is our 25 RPM reference spectrum. The averaged and final reference spectrums are shown in Figure 10.12.

Figure 10.12: Average and Reference (Smoothed) Fourier Spectrums

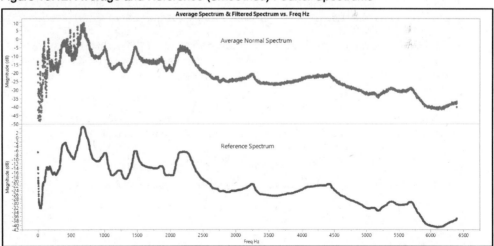

A procedure very similar to that outlined in Figure 10.11 is performed as the data streams in. The only difference is that we average Fourier spectrums over an interval of 20 seconds rather than a minute. We then smoothed these average spectrums with a median filter and compared them to the reference spectrum.

Figure 10.13: Averaged Fourier Spectrum (20 secs.) Versus Reference (Gray=Normal, Black=Intervention)

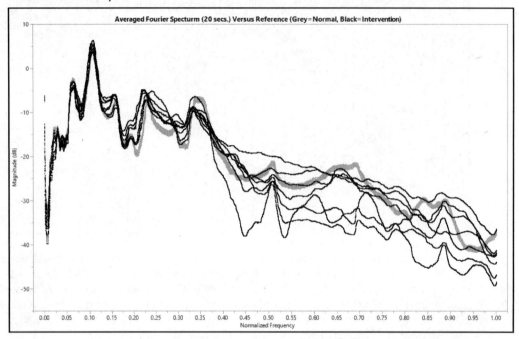

Figure 10.13 shows a representative sample of the averaged spectrums of the streaming data along with the reference spectrum. Twenty-six normal spectrums are shown in gray (all overlap and hide the reference spectrum) and seven spectrums after the intervention (black).

The reference spectrum is difficult to see because so many of the normal data spectrums lie directly on top of it. Of all the normal spectrums, only one did not lie on the reference spectrum (not shown here), whereas most if not all the dark non-normal spectrums lay clearly away from the reference.

To effectively monitor the differences in spectrum on streaming data, we need a useful measure of distance from the reference spectrum. Here we introduce two measures:

- Mean Absolute Distance: $\boldsymbol{D_{mean}} = \boldsymbol{Mean}(|\boldsymbol{S_{in}} - \boldsymbol{S_{ref}}|)$
- Hausdorff Distance: $\boldsymbol{D_{Hausdorff}} = \boldsymbol{D_H}\,(\boldsymbol{S_{in}}, \boldsymbol{S_{ref}})$ where $\boldsymbol{D_H}(\boldsymbol{X},\,\boldsymbol{Y}) =$

$$\boldsymbol{max}\left(\sup_{x \in X}\inf_{y \in Y}\boldsymbol{d(x,y)}\,,\sup_{y \in Y}\inf_{x \in X}\boldsymbol{d(x,y)}\right)$$

where $\boldsymbol{S_{in}}$ and $\boldsymbol{S_{ref}}$ are points on a streaming spectrum and the reference spectrum, at a given frequency, respectively. (See Figure 10.14.)

Figure 10.14: Calculating Distance from Reference Spectrum

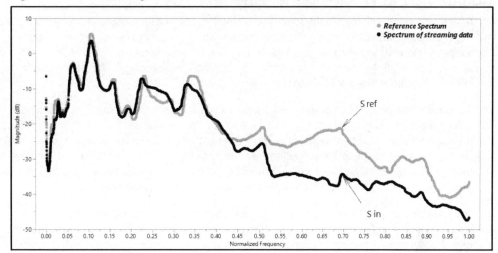

Note that the Hausdorff distance is simply the largest of all the distances from a point on one spectrum to the closest point on the other.

Figure 10.15 shows the results of the distance measures for all the 25 RPM data.

Figure 10.15: Distance Measure Results

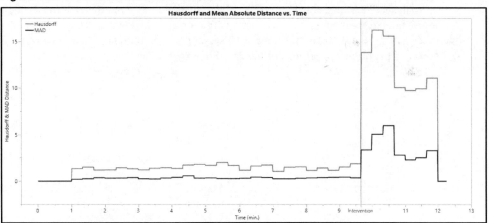

The Hausdorff distance is shown in green (light shade) and the mean absolute distance in black. Both distance measures detect the change in model state only a half a second after the intervention! As with the time domain analysis and the digital filter, we successfully implemented the entire workflow just outlined on SAS Event Stream Processing. Because the streaming version shows a figure identical to Figure 10.15, we do not show it here.

Monitoring the Whole Fourier Spectrum by Segments

In practice, once a change in the spectrum is flagged, the machine operator would likely want to investigate further to figure out which frequencies or frequency bands were the likely source of the shift in the spectrum. Rather than whole-spectrum monitoring at once as we did above, they might be interested in segment-by-segment monitoring of the whole spectrum. We illustrate how this can be done.

To monitor the entire Fourier spectrum by frequency band segments, our first task is to determine the frequency segments. There are several ways to do this. First, we can perform the segmentation based on frequencies most relevant given the motor design. Second, we can look at the average spectrum under "normal" or fault-free, stable conditions and segment it using common sense. Because we are not vibration or machine experts, we decided to choose the second way and use statistics to help us come up with the segments.

We tried segmenting the "normal" Fourier spectrum based on several representations of it (such as power, magnitude, cumulative power over frequency, and cumulative magnitude over frequency). The segmentation based on cumulative magnitude gave us what seemed to be the most logical segmentation results. First, sections of the spectrum with higher power were largely in the same segment. Further, the segmentation results were quite similar across the three different speed experiments. Lastly, our ability to detect the change in model state was similar no matter which experiment's segmentation results were used. Thus, we segment cumulative (over frequency) magnitude.

To segment the cumulative magnitude, we used the SEGMENTATION function in the Time Series Analysis (TSA) package in the TSMODEL procedure (Leonard and Trovero 2014). We approximated the cumulative magnitude curve for the 50 RPM experiment using 13 linear segments. Figure 10.16 shows these 13 segments on both the cumulative magnitude and power of the normal spectrum for the 25 RPM experiment.

Figure 10.16: Frequency Segments of Cumulative Magnitude (dB) 25 RPM Experiment

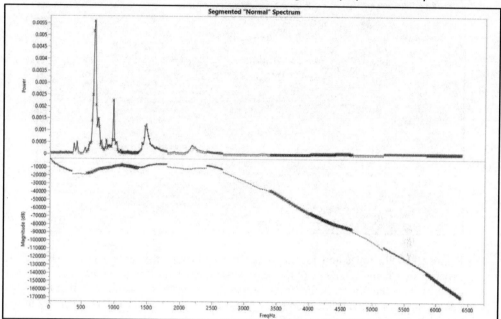

After the frequency segments to monitor were identified, we began our offline training. The offline training here begins like the offline training for monitoring the whole spectrum at once. (See Figure 10.17.)

1. We began by performing STFT on the "normal" data using a window size of 4096 observations and then skipping ahead 2048 observations.
2. We then calculated a moving average of 125 STFT windows (20 seconds). For each of these 20-second average spectrums, we calculated the average spectrum magnitude in each of the 13 frequency segments.
3. Finally, we evaluated how the magnitude in each segment varies across time under the "normal" condition.

The small graph in the bottom right of Figure 10.17 shows the relative stability of the magnitude of the averaged spectrums in the 13 segments over time under the "normal" condition. Upper and lower bounds are easy to establish, using, for example, a +/- 3-sigma rule for each segment.

Figure 10.17: Offline Training: Monitoring Frequency Segments

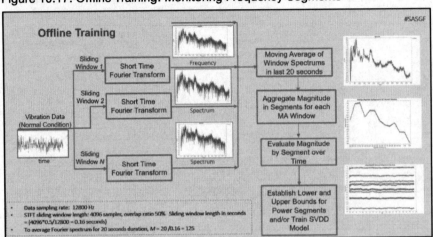

Figure 10.18 illustrates how the magnitude in these segments varies over the whole 25 RPM experiment. Changes in many of the segments immediately following the intervention are obvious. Note that the transition from normal (to the left of the vertical line) to the right of the vertical line is gradual as the samples in between initially (20 seconds) have a mix of the "normal" and post-intervention data.

Figure 10.18: Average Magnitude in 13 Frequency Segments Over Time

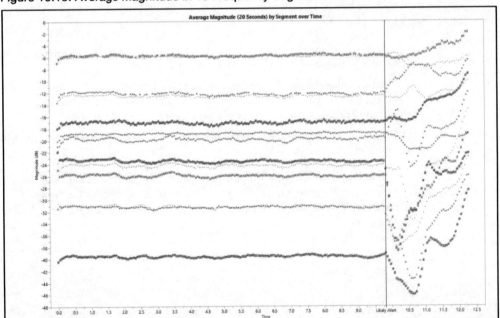

Figure 10.19 shows monitoring results from segment 11 (5170–5852 Hz) using the +/- 3-sigma rule. An alert is triggered in slightly more than seven seconds following the intervention.

Figure 10.19: Average Magnitude in 5170-5852 Hz Segment with +/- 3 Sigma Bounds

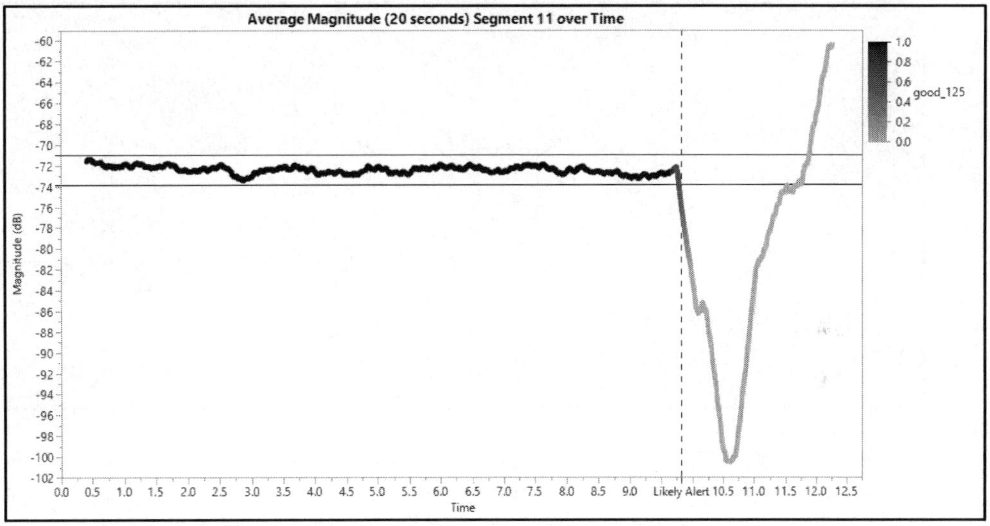

All segments can be monitored together just like when we monitored the whole spectrum earlier.

If we change the rotational speed of the motor, we expect the Fourier spectrum to change. Thus, the average magnitude in the various segments also changes. This does not render the approaches invalid. It just means that you have to compute a new reference spectrum and/or determine new bounds on the frequency segments. This is a relatively simple thing to do.

In order to avoid having to change reference spectrum and/or bounds as the machine conditions change, one possible approach is Support Vector Data Description (SVDD) (Tax and Duin 2004). SVDD can theoretically handle multiple "operating modes." Briefly, SVDD is a one-class classifier. In our case, the one class is the "normal" operating condition and in our experiments, normal includes running the machine at 25, 35, and 50 RPM (multiple operating modes) when everything is "normal." Intuitively, SVDD finds a minimum radius boundary around the data characterizing the normal condition(s).

Then, as additional data is generated, the model assesses whether the new data are close enough to the "normal" operating condition description. SVDD can handle multivariate data and the boundaries around the data can be flexible when a kernel function is used. To see this, consider Figure 10.20 where we illustrate SVDD results for a system that is assumed to have three different operating modes and only two sensors.

- The normal operating data are shown in the left graph.
- The SVDD results are shown in the right graph.

- The areas of normal operation as characterized by the SVDD model are shaded in dark gray. Any new data streaming in that lie outside these gray areas would be considered NOT "normal."

Figure 10.20: SVDD Under Multiple Modes of Normal Operation

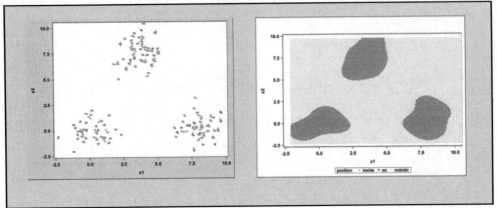

In theory, a single SVDD model could capture the normal condition of our motor running at multiple speeds. How well the model works in practice depends on whether the non-normal modes at the different speeds are also outside all the normal modes (at all speeds).

To train an SVDD model for our three experiments, we used 75% of the normal data for each speed. The value of Gaussian bandwidth parameter plays an important role in getting a good training data description. We used the SVDD procedure available in SAS Visual Data Mining and Machine Learning. The SVDD procedure supports Mean (Chaudhuri et al. 2017), Mean2 (Liao et al. 2018) and Trace (Chaudhuri et al. 2018) criterion for bandwidth selection. We use the Trace bandwidth selection criterion in this analysis. We then scored the model on all three data sets to assess model performance to see how quickly it can find where we intervened at each speed. Note that since we model all three experiments together, we just have one critical value for all the speed experiments.

Figure 10.21 shows the performance of our SVDD model for the 25 RPM experiment. It alerts that there is a change from the normal condition 7.6 seconds after the intervention. Notice that the alert comes when the spectrums being averaged contain a mix of the "normal" and "changed" condition!

Figure 10.21: SVDD Results 25 RPM Experiment

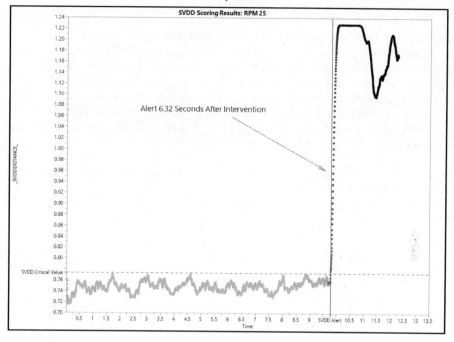

The SVDD results for the other two experiments are shown in Figure 10.22. The results are like those in the 25 RPM experiment. These results suggest that if you train an SVDD model using normal conditions from all speeds at which the motor is to be run, the model should be able to detect changes in normal conditions.

Figure 10.22: SVDD Results 35 RPM (Left) and 50 RPM (Right)

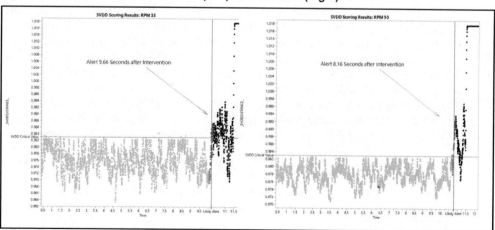

Whether the model remains successful for other non-normal conditions depends on how different from the normal operation at any speed these non-normal conditions are. How could one assess whether this was true or not (presuming other fault data are available)? All we would need to do is to visualize our 13-dimensional data!

Fortunately, there have been some interesting developments in tools designed to help visualize multidimensional data. The TSNE procedure in SAS VDMML implements one such method of visualizing multidimensional data – t-distributed Stochastic Neighbor Embedding (Van der Maaten and Hinton 2008). Examining the 13 variables used to train and score our SVDD model help us understand the performance of the SVDD model shown above. Recall these 13 variables describe the average magnitude of the Fourier spectrum in the different frequency segments. We use the TSNE procedure to visualize these 13 variables in two dimensions. The results are shown in Figure 10.23.

Figure 10.23: t-SNE Results by RPM (25 Dark, 35 Dark Gray, 50 Light Gray) on Left; Normal (Gray)/Not (Dark) on Right

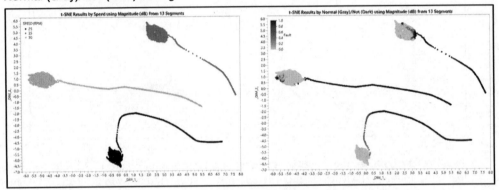

The left panel in Figure 10.23 shows the two-dimensional visualization of our data by speed: dark = 25 RPM, dark gray =35 RPM, and light gray =50 RPM. Interestingly, most of the data from each of the RPM experiments lie in an almost circular data cloud. But then there are some odd tails. A look at the right panel shows exactly what the tails are about. They are the data points post-intervention! Although the data points are not completely visible, you can see that some of the post-intervention data for the 35 and 50 RPM experiment do lie in the "normal" data cloud – explaining the slightly worse SVDD results for these experiments. Although we cannot show an animation here, if you animate the above graphs by time, you can see an almost circular motion in the dense circles. Then, once the intervention happens, the data shoots away from the "normal" area. The TSNE procedure provides valuable insight into what the normal or stable, fault-free motor conditions look like for different motor speeds and why the SVDD model works well for the intervention in these experiments.

Once again, we can do all the above real-time analysis using SAS Event Stream Processing with data streaming at 12800 reads per second. Display 10.2 is a screen capture of the Streamviewer of the Frequency Segment Monitoring and the SVDD model scoring results as the data stream in. The SVDD model is trained offline and the incoming data scored using an ASTORE.

Display 10.2: ESP Streamviewer: Frequency Segment Monitoring and SVDD Score Results

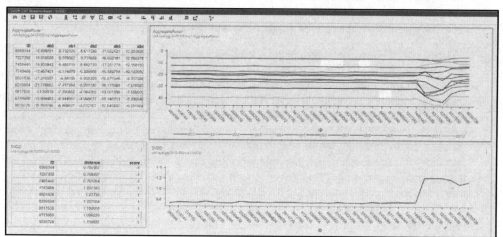

Conclusion

This chapter explored the effectiveness of some of the more advanced SAS analytical tools to analyze sensor data streaming at high sampling rates. Given the large volume of data needed to study machine vibrations and the importance of vibration analysis in assessing the health of high-valued machines for condition-based maintenance, we focused on vibration data.

We used a variable speed, three-phase Induction Squirrel Cage Motor equipped with three accelerometer vibration sensors. Using this machine, we ran three experiments with the motor running at 25, 35, and 50 revolutions per minute (RPM). First, the motor was run in its usual stable state or under "normal" conditions, and then we intervened to slightly change the adjustment on the motor shaft. We then proposed several methods to analyze the vibration data with the goal of early detection of the changed state due to the intervention.

Initially, we focused on time domain analysis, and we monitored the Root Mean Square of the Amplitude (RMSA) and the kurtosis of the raw vibration sensor series using moving windows of data over time. We detected the change in motor state slightly more than one minute after the intervention in real-time using SAS Event Stream Processing. Next, we illustrated how we could detect the change in model state much faster with a combination of frequency and time domain analysis. We demonstrated how to isolate the frequency band of interest using the new Butterworth digital filter and then used the ratio of RMS amplitude of the filtered time series relative to the original signal to detect change. This ratio was able to detect the machine change of state in under eight seconds. This vibration analysis technique, demonstrated in real time in SAS Event Stream Processing, is very useful in cases where analysts know exactly which frequencies need to be monitored based on the machine design and its operating parameters.

Next, we switched entirely to the frequency domain. Here we proposed a method of monitoring changes in the entire Fourier spectrum using Short-Time Fourier Transform (STFT) and a novel, relatively simple distance measure (Hausdorff). Both STFT and the distance measures were easy to implement on streaming data in SAS Event Stream Processing, and this method detected the change in model state under one second after the intervention. This method has the slight inconvenience of requiring recalibration of the reference spectrum whenever the motor condition (speed or otherwise) changes.

To help analysts understand the specific frequency segments that seem to be changing, we proposed another method of monitoring the full Fourier spectrum, but now we monitored the spectrum in frequency segments. We proposed several techniques for segmenting the spectrum, including using the SEGMENTATION function in the Time Series Analysis (TSA) package in the TSMODEL procedure. Once the segments were determined, the monitoring technique used STFT. The results from STFT were aggregated by segment to greatly reduce the data required in analysis. We demonstrated how to monitor the average magnitude of the frequency in each segment using simple threshold limits and/or train an SVDD model using the "normal" operation data and then score on incoming data. We showed how both the individual segment monitoring and SVDD model can be implemented in real time in SAS Event Stream Processing and how they both detect the change in motor state very quickly. A nice advantage of the SVDD model is that the same model can be used to monitor the motor running at multiple speeds if normal operating data at each of these speeds are used in training.

The surprisingly effective performance of the SVDD model at all motor speeds examined in this study made us curious about how disparate the normal-state data clouds were at the different speeds and where the data from the altered model state were relative to the normal states. To help us understand what was happening in the 13-dimensional space occupied by our data, we used a new TSNE procedure at SAS that enables you to look at high-dimensional data in two or three dimensions. The results showed clearly why SVDD worked so well. An obvious extension of this work would be to expand the experiments to include more speeds and include various load conditions and other faults to see whether the success of the current analysis generalizes under more scenarios.

Clearly, you can use very advanced analytical techniques with SAS Event Stream Processing to successfully monitor machine health with data streaming online at high sampling rates. Applying this level of intelligence right at the edge is essential to effective condition monitoring.

References

Chaudhuri, A. et al. 2017. "The Mean and Median Criterion for Automatic Kernel Bandwidth Selection for Support Vector Data Description." In *2017 IEEE 17th International Conference on Data Mining Workshops* (ICDMW). Piscataway, NJ: Institute of Electrical and Electronics Engineers.

Chaudhuri, A. et al. 2018. "Sampling Method for Fast Training of Support Vector Data Description." In *Proceedings of the 2018 Annual Reliability and Maintainability Symposium (RAMS)*. Piscataway, NJ: Institute of Electrical and Electronics Engineers.

Chaudhuri, A. et al. 2018. "The Trace Criterion for Kernel Bandwidth Selection for Support Vector Data Description" Available https://arxiv.org/abs/1811.06838.

Leonard, M. and Trovero, M. 2014. "How to Separate Regular Prices from Promotional Prices." *Proceedings of the SAS® Global Forum 2014 Conference*. Cary, NC: SAS Institute Inc. Available https://support.sas.com/resources/papers/proceedings14/SAS212-2014.pdf.

Liao, Y. et al. 2018. "A New Bandwidth Selection Criterion for Using SVDD to Analyze Hyperspectral Data." In *Proceedings of the 2018 Defense and Security Symposium*. Bellingham, WA: SPIE.

Oppenheim, A.V. and Schafer, R.W. 2009. *Discrete-Time Signal Processing,* 3rd ed. New York: Pearson.

Randall, R.B. 2011. *Vibration-Based Condition Monitoring: Industrial, Aerospace and Automotive Applications.* West Sussex, UK: John Wiley and Sons, Ltd.

Sejdić, E, Djurović, I., and Jiang, J. 2009. "Time-frequency feature representation using energy concentration: An overview of recent advances." *Digital Signal Processing.* 19 (1): 153–183.

Tax, D.M.J., and Duin, R.P.W. 2004. "Support Vector Data Description." *Machine Learning* 54 (1):45–66.

Van der Maaten, L. and Hinton, G. 2008. "Visualizing Data Using t-SNE." *Journal of Machine Learning Research* 9:2579–2605.

About the Contributors

Anya McGuirk is a Distinguished Research Statistician in the Advanced Analytics Division at SAS where she currently focuses on analytics related to the Internet of Things. She joined SAS in September 2004. Prior to that, Anya was a Full professor at Virginia Tech with a joint appointment in the Departments of Agricultural and Applied Economics, and Statistics.

Yuwai Liao is a Senior Research Statistician Developer in the Internet of Things (IoT) department at SAS Institute. She received her PhD in Electrical and Computer Engineering from Duke University. Her research interests includes digital signal processing, signal detection, computer vision, and machine learning.

Byron Biggs is a Principal Research Statistician in the Internet of Things Division at SAS, where he specializes in analytics for high-frequency streaming data and high-performance computing. Prior to SAS, Byron worked in high-performance distributed computing for software companies and in physics/software consulting. Byron holds a PhD in Physics from the University of Virginia and a BSE in Electrical Engineering and Computer Science from Princeton University.

As a Principal Research Statistician Developer, **Deovrat Kakde** performs research and development of techniques useful for analyzing sensor data. He has published extensively and holds multiple patents. He has more than 20 years of experience in various industries such as transportation, software development, and engineering.

As a Software Developer at SAS, **Joseph Costin** works on the latest in data mining and automated machine learning applications, which are used by customers to solve problems spanning all industries. Joseph holds a bachelor's degree in Computer Science from North Carolina State University.

Chapter 11: Analytics with Computer Vision on the Edge

By Juthika Khargharia and Hamza Ghadyali

Introduction

Imagine that every camera is a smart AI-equipped camera. Think of the business value that such cameras would add to the manufacturing, retail, and utilities industries. But this possibility does not need to be left to the imagination. Now that powerful hardware in edge devices has become more affordable, those devices can be used in combination with the latest advances in the SAS platform. We now can create intelligent computer vision systems by building complex deep learning models with SAS Visual Data Mining and Machine Learning (VDMML) on SAS Viya, integrate them with other SAS analytics tools, and deploy those models on edge devices to score streaming data using SAS Event Stream Processing. With that framework in place, we can integrate the newest data into our analysis, and make decisions in *real time.*

This chapter explores how you can use computer vision to solve problems with real business value. It shows how you can build computer vision models with deep learning to solve previously unsolvable problems through examples from specific applications. It also shows the advantages of deploying these models for real-time analytics and provide some resources for getting started in this exciting field.

Computer Vision with Deep Learning

How does a computer "see" an image? Why has it been a challenge to perform analytics on images? To answer these questions, we first must think of an image as a grid of colored dots (pixels). Colors can be represented as a mixture of three primary colors, such as **red**, **green**, and **blue** (RGB representation). Thus, three numbers are needed to represent a color for a single pixel. If you have a 12-megapixel camera, then each image is represented by $3*12 = 36$ million numbers. If colors are stored as 8-bit, then each of these numbers takes on a value between 1 and $2^8 = 256$ (in practice, the range is 0 to 255). Put another way, an image is represented numerically by three matrices (or 2D arrays). Each matrix is called a *channel*, and there is one channel for each of the primary colors red, green, and blue. (See Figure 11.1.)

Figure 11.1: Image of a Cat Represented Numerically as Three 2D Arrays

Now imagine zooming into a 5x5 patch of pixels on that image. In Figure 11.2, the color of the square in the top left corner is represented by three numbers (161 for red, 36 for green ,7 for blue). Seen on the right, there are three 5x5 arrays that explicitly show the color components for each square in the left image.

Figure 11.2: Zooming into a Patch of Pixels

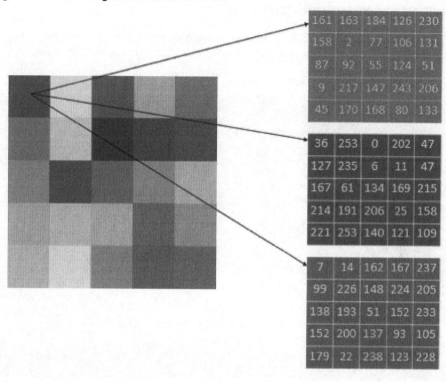

A typical movie or television show might stream at 30 images every second. This volume can leady to truly big data, which presents two challenges. First, there is managing such a large amount of data that is streaming in at a rapid rate. Second, there is identifying objects of interest, which involves complex relationships between groups of pixels.

Simple rules to classify and track objects do not exist. Deep learning can be used to develop a model that learns complex rules with minimal human intervention. Deep learning makes image classification, object detection, and object tracking much easier. This does not mean that traditional computer vision and image processing techniques are obsolete. Those techniques continue to play an important role in making models robust to the types of variation that is seen in true production environments where the models will ultimately need to perform.

The basic pipeline for building computer vision problems usually involves taking a supervised learning approach. First image or video data is collected. Then, a subset of that data is selected for labeling that is used for training the model offline. A model's performance can be validated with the remaining data. After a model is performing well, it is deployed. In most use cases, real-time deployment has the most value.

Powerful models that leverage cutting-edge analytics can be built using SAS VDMML and are best deployed for real-time applications with SAS Event Stream Processing. When you combine these two products with the rest of the SAS platform, further advanced analytical models can be built, and after deployment, both data and models can be managed and monitored for continued performance.

Figure 11.3 shows the end-to-end computer vision pipeline starting with model training in SAS Viya and real-time deployment of these models in SAS Event Stream Processing.

Figure 11.3: Model Training and Deployment

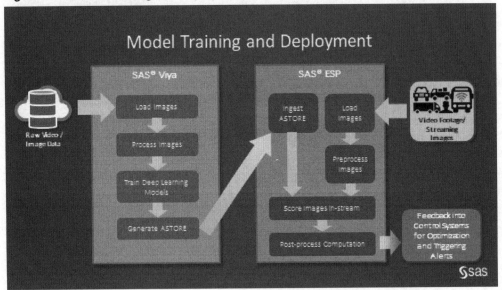

To achieve the pipeline described in Figure 11.3 using SAS software, one can specifically take advantage of DLPy in SAS Viya for model training and SAS ESP for model deployment. DLPy is a high-level Python package for the SAS Deep learning features available in SAS Viya. DLPy is designed to provide an efficient way to apply deep learning methods to image, text, and audio data.

Example Code 11.1 shows a Python code snippet from DLPy that generates and saves an ASTORE file from a pre-trained VGG16 model.

Example Code 11.1

```
model_vgg16.deploy(path='<path>', output_format='astore')
```

An *analytic store file* (ASTORE) is a binary file that contains the state of a trained predictive analytic model. A key feature of an analytic store file is that it is transportable between platforms and hence can be easily ingested by SAS Event Stream Processing for real-time deployment.

Figure 11.4 shows the ingestion of the ASTORE file using the Model Reader window in SAS Event Stream Processing Studio to score new images. A Model Reader window (model_reader) receives requests from a Request window (w_request), fetches the specified model using the request information, and publishes the model event to the Score window (score) for scoring.

Figure 11.4: Process Flow Showing Ingestion of an ASTORE File and Scoring of New Images

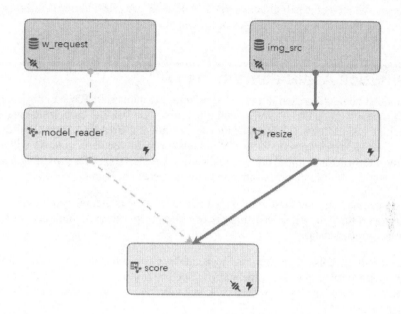

Example Code 11.2 shows the XML code underlying the Model Reader window in SAS Event Stream Processing Studio.

Example Code 11.2

```
<window-model-reader name='model_reader' model-type='astore'/>
```

Advantages of Real-time Analytics on the Edge

In the context of computer vision, there are several additional advantages of building and deploying models that can process image and video streams in real time. From a computational point of view, there are the advantages of smarter and more efficient data management. Instead of streaming in large quantities of data into a storage device for batch analytics at a later date, real-time processing reduces the network burden and processes the data as it arrives, and rules and models can be designed to store only the most important data for future use.

Real-time analytics using computer vision also has the benefit of augmenting human effort by running side-by-side with skilled workers. Real-time analytics reduces the delay between data collection and data analytics. This reduction is crucial for applications such as the detection of suspicious activity in surveillance footage or the detection of defects in products during manufacturing, or the other use cases discussed next.

Luckily, we can reap the rewards of these advantages by deploying computer vision models with SAS ESP. SAS ESP is optimized to run on affordable Nvidia GPU hardware for real-time model scoring. The outputs of the deployed models can be monitored with SAS ESP Streamviewer, where you can build custom dashboards to provide additional insights augmenting the plain footage and images of the camera.

Computer Vision Applications in the IoT

Applications of computer vision in the IoT span multiple industries. We summarize a few applications below in industries such as manufacturing, retail, health care, insurance, transportation, and utilities. These applications show how the breadth and depth of the SAS platform allow us to build a wide variety of models for many use cases, and we will see how we transform unstructured image and video data into structured metrics providing actionable analytics to deliver value, support innovation, and complement human skills.

A key advantage to building these models with deep learning is that we can build robust models using multiple inputs, such as images or videos from cameras combined with sensor data, or other process data.

The value for all these use cases is in real-time deployment where streaming data (images, video, sensors) is analyzed with sophisticated models providing actionable intelligence that can automatically send out alerts or feed back into control systems in real time, which is powered by SAS ESP.

Manufacturing

Downtime Reduction in Production Line Monitoring

Deep learning object detection models can be used in manufacturing to avoid downtime in production lines. By robustly tracking products as they move through the production line, we can monitor inefficiencies in the flow; detect defects; or detect, predict, and prevent costly collisions. Collisions can result in product damage, or worse, a shutdown of the continuous production line, resulting in major losses in productivity.

Figure 11.5 displays the result of a computer vision model that is detecting objects, tracking them with IDs, and determining relative positions and velocities of objects. The output of the computer vision model can then be further processed with models built using the broad advanced analytics capabilities afforded by the SAS platform to trigger an alert in the event of a detected defect, irregular flow, or a collision.

Figure 11.5: Production Line Monitoring

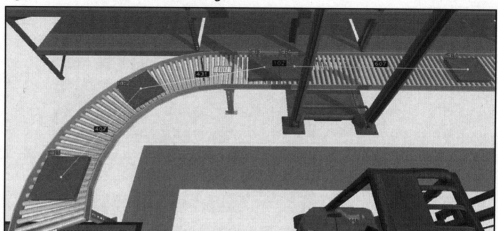

Defect Detection in the Semi-conductor Industry

In order to improve yield and optimize productivity in semiconductor devices, it is necessary to implement reliable inspection methodologies. Vision inspection technologies have been developed to analyze images of wafers for defects. Defect detection is an important part of the wafer fabrication process. Defects can be classified into different categories based on specific patterns, geometries, wafer scratch, and so on.

Figure 11.6 shows an example of such defect categories. After the defect is detected, it needs to be classified to one of several defect categories; this enables correction of the fabrication process.

Figure 11.6: Example of Wafer Defects

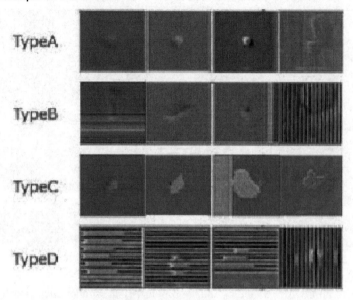

(Source: Kazunori Imoto, Tomohiro Nakai, Tsukasa Ike, Kosuke Haruki, Yoshiyuki Sato. 2019."A CNN-Based Transfer Learning Method for Defect Classification in Semiconductor Manufacturing," *IEEE Transactions on Semiconductor Manufacturing*, vol. 32, no. 4, pp. 455-459.)

Deep learning models such as convolutional neural networks can be trained against historical image data to learn the defect features accurately. The trained model is then applied to new incoming defective image data to categorize them into one of the several defect types. It is important to note that the defects themselves might evolve over time, and new defects can emerge. Although it is important to re-train the models using new data frequently, unsupervised learning models can also go a long way toward identifying new classes of defects. SAS provides a full stack of machine learning, deep learning, and statistical models that can be used to train image data. The trained models can then be used to classify new defects in real-time thereby enhancing the productivity of the operator.

Quality Inspection in Surface Mount Technology (SMT) Printed Circuit Boards (PCBs)

To enhance the quality of PCB production, it is imperative to perform robust inspection of the boards. Good quality SMT inspection machines are often used to look at missing components (for example, capacitors, resistors, and so on), inspection of skewed components, classification of components into different categories, and so on. Advanced Optical Inspection (AOI) machines as well as Advanced X-ray Inspection machines (AXI) are often used in conjunction to captures images of the boards in the optical and X-ray wavelengths respectively. This enables the operator to analyse not only the visually available components but also the solder joints. With X-ray technology, one can obtain a direct view of the solder joints; data from this could be used for analysis of joint quality, Head-in-Pillow (HiP) defects, and so on.

Figure 11.7 shows sample images taken from an AXI machine. It highlights issues with the solder join such as too much solder, misplaced solder, or errant solder balls.

Figure 11.7: Sample Images from AXI Machines

(Source: Jonathan Titus. 1999. "X-Rays Expose Hidden Connections." *Test & Measurement World*: 28-35.)

The data from known defects and non-defects are captured to train convolutional neural networks. After the defects are learned by the model, the scoring logic is sent to a real-time inference engine software that is installed on the inspection machines. This enables real-time scoring of the image components as they are scanned. It also enhances operator efficiency by relieving operators of their manual visual inspection work. One of the challenges faced is collection of enough data that represents defects so that deep learning models can be trained effectively. A possible work-around to this challenge is implementation of several image augmentation techniques that SAS provides out-of-the-box to increase the size of the training data set. Models can be continually improved as new defects are added to the defects database.

Government

Smart City Applications

A *smart city* is defined as an urban area where sensor data is collected from various connected devices and assets in the city for efficient allocation of resources and assets to bring overall welfare to the citizens. While use cases around smart cities abound, here we focus on one such case: monitoring and management of traffic flow and transportation systems. Many cities around the world have invested heavily in installing CCTV cameras at several busy intersections in their cities. As part of the smart city initiatives, city officials want to apply machine learning and deep learning on the video feeds coming off these cameras. The city officials are interested specifically in several metrics such as vehicle classification at traffic junctions; counts based on vehicle type; real-time speed monitoring; traffic congestion analysis based on time of the day, day of the week, and holidays; monitoring vehicles that violate rules such as wrong turns; U-turn violations; and so on. The city command and control center are interested in collecting these key metrics from multiple video feeds, aggregating the data, and monitoring real-time reports.

This use case is solved using a two-pronged approach using SAS multi-phase analytics. The first part involves training deep learning object detection models that learn to differentiate between vehicle types, pedestrians, and so on. The second part is concerned with model deployment. Once we have a fully trained model, it can be deployed on a GPU-enabled edge device that will score new incoming data against the prebuilt deep learning model. Using edge software, one can build out the logic for lane violation using geofencing in real-time, speed computation, monitoring counts of different vehicle types based on user-defined conditions.

Figure 11.8 shows a real-time dashboard in SAS Event Stream Processing Streamviewer displaying real-time values of traffic monitoring metrics such as vehicle identification, vehicle counts over a certain time frame, and speed computation.

Figure 11.8: SAS ESP Streamviewer

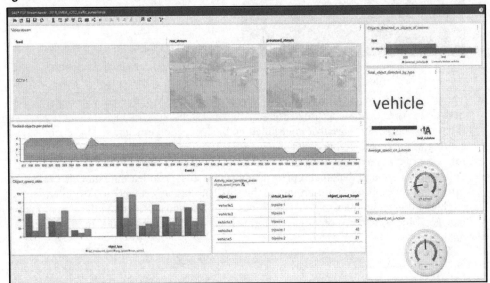

Data Management for Video Surveillance

Although video surveillance data can be used for a variety of analysis, it does come with a cost associated with video data storage. With the emergence of high-definition cameras and longer retention times, many government agencies are dealing with the cost of storing this data. Because video data is unstructured in nature, it requires special storage to be cost-effective. One way to combat this challenge is post-processing of video data to significantly reduce the amount of storage. Not all data might be valuable to store. Determining how much data to transmit and store is a concern surrounding many IoT applications.

One method for reducing video footage data is called *robust principal component analysis* (RPCA), an algorithm supported by SAS under SAS VDMML. RPCA decomposes an input matrix into the summation of a low-rank matrix and a sparse matrix. The robustness of the method comes from its ability to handle anomalies. Anomalies are captured within the sparse matrix and can be used to examine potentially abnormal behavior within the data set. RPCA is a great option for dimensionality reduction in video data.

In Figure 11.9, the original matrix captures the raw video footage (left panel). The low rank matrix captures the static background (middle panel) and the sparse matrix captures the moving background (right panel). In many cases, storing the moving foreground footage can provide enough data for analysis and thereby reduce storage.

Figure 11.9: Decomposition of Video Content

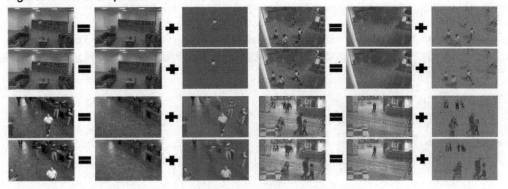

(Source: Zhou, Tianyi. 2011. "GoDec: Randomized Low-rank & Sparse Matrix Decomposition in Noisy Case."
Tianyi Zhou's Research Blog. https://tianyizhou.wordpress.com/2011/04/20/learn-low-rank-sparse-
structures-via-randomized-alternating-projections/)

Transportation

Loose Ballast Detection in Railroads

Have you ever wondered why trains pass through railroad tracks that are covered with jagged little stones? The reason is that these stones are the most efficient and cost-effective way for a robust foundation for the train tracks. They are called track ballast and they seek to keep the tracks in place. Railway tracks can change over time due to weather conditions, vibrations, ground movement, and weed growth that makes the tracks unstable. The train ballast protects them from such events. Figure 11.10 compares normal ballast relative to loose ballast (less gravel in the train tracks) within the red highlighted area of interest.

Figure 11.10: Normal Ballast (Left) Versus Loose Ballast (Right)

A computer vision application surrounding this use case has to do with loose ballast detection. The objective is to identify loose ballast and monitor track health conditions in real time using cameras mounted on the trains. These cameras are continuously capturing video of the train tracks. A deep learning model trained on historical video data learns to identify the features associated with the loose ballast condition. This model is then applied to real-time video feeds to monitor the track health alongside other structured metadata captured with the video. Figure 11.11 shows SAS Event Stream Processing Streamviewer monitoring track health along with geo-location in real-time. Other great applications in the field of transportation involve monitoring traffic, identifying vehicles that are driving the wrong way, and spotting detritus or other adverse conditions on roads or highways.

Figure 11.11: SAS Event Stream Processing Streamviewer Monitoring Track Health

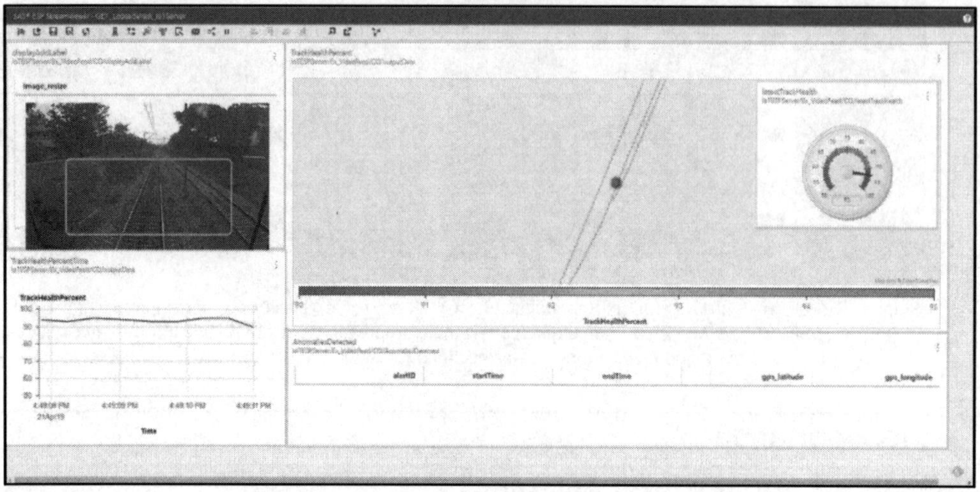

Health Care

Biomedical Image Analysis

According to GE Healthcare, "Hospitals are producing 50 petabytes of data per year. A staggering 90 percent of all health care data comes from medical imaging" (GE Healthcare 2018). Biomedical image processing intersects with multiple fields such as computer science, machine learning, image processing, medicine, and other fields. Typically, radiologists would interpret biomedical images to diagnose medical conditions. However, with the growing volume of imaging studies, a radiologist's workload is extremely demanding.

Here we discuss how SAS tools can potentially be used to alleviate a radiologist's time-consuming and demanding task. SAS supports various health and life sciences use cases ranging from management of clinical health care data to taking advantage of medical images along with statistical, data mining, text analytics, and optimization techniques for better clinical diagnosis. SAS can directly read the commonly available biomedical imaging formats such as Digital Imaging and Communication in Medicine (DICOM), Neuroimaging Informatics Technology Initiative (NIFTI), nearly raw raster data (NRRD), and so on. One can then take advantage of the biomedical image pre-processing, segmentation, visualization, and deep learning frameworks in SAS for various use cases.

Below is an example of lung nodule classification on image data provided by Society of Photographic Instrumentation Engineers (SPIE), American Association of Physicists in Medicine (AAPM), National Cancer Institute (NCI), and TCIA. Figure 11.12 shows an example of lung nodule classification.

Figure 11.12: Lung Nodule Classification Based on Metrics Derived from Engineered Features (Left) and Deep Learning Analysis (Right)

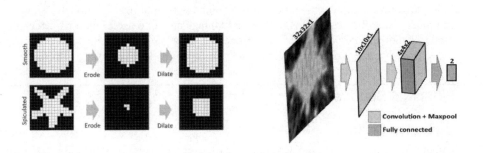

Two approaches were used in Figure 11.12. The left panel shows how metrics derived from engineered features can be used for classifying a benign versus a malignant nodule. Subsequent image processing steps of dilation and erosion preserves the shape of a round (possibly benign) nodule. Similar image processing steps on a spiculated (possibly malignant) nodule does not preserve the initial geometry of the nodule. Figure 11.12 (right) shows deep learning analysis applied to the sample benign and malignant image data, giving a 10% misclassification rate (Vadakkumpadan and Sethi 2018).

Utility

Vegetation Encroachment Monitoring on Power Lines Using Drones

Management of overgrown trees and vegetation that interfere with transmission lines is perhaps one of the most expensive operational costs for distribution of electricity. These vegetation encroachments are monitored through time-consuming visual inspections periodically, but it has proven neither to be accurate nor cost effective. Energy and utility companies are now starting to explore new technologies such as Unmanned Aerial Vehicles (UAVs) also known as drones.

For example, Duke Energy is using drones to conduct infrared equipment inspections, survey storm damage, and inspect tall structures (Wells 2018). Drones have also brought about advantages in terms of safety. Since drones are unmanned, no one needs to be inside an aircraft flying at low altitudes parallel to transmission lines. Drones can cover large distances in a single flight and provide detailed and accurate aerial images of transmission lines, surrounding vegetation, and other structures. This could lead to automation of a very expensive traditional manual process.

This use case can be tackled in three different stages, each of which requires detailed analytics drawn from computer vision, traditional machine learning, and operations research. The first stage typically relies on analysis of the images collected by the drones. One can build robust object detection models for classification of vegetation into several species. The trained models can be deployed on the drone equipment to classify these vegetation species in real time. In stage two, analytical models can be built to predict growth rates of these species.

Stage three is concerned with joining the vegetation growth forecast models with maintenance and cost information to provide vegetation management intervention in terms of routing and allocation of resources.

This shows how computer vision can be combined with optimization models, where computer vision first transforms unstructured image data into useful metrics, which then are used to optimize complex systems and processes. This integration with SAS Optimization shows the advantages of the breadth of analytic capabilities of the SAS platform.

Financial Services

Insurance Claims

Effective insurance claims processing is key to insurers' operational efficiency, because they have only a short window within which to process these claims. Computer vision is starting to become inevitable for any insurer wanting to have detailed information about insured property.

By using the SAS Scripting Wrapper for Analytics Transfer (SWAT) package and SAS DLPy (a high-level Python library for the SAS deep learning features available in SAS Viya), a user can quickly perform image processing and SAS deep learning to classify defects. The SWAT package is a Python interface to the SAS Cloud Analytic Services (CAS) engine. It includes the ability to call CAS actions and process the output in various ways. These range from simple (calling CAS actions as Python methods and getting a dictionary of results back) to complex (invoking CAS actions in multiple sessions and handling server responses from them directly).

Figure 11.13 shows image preprocessing techniques being applied to damaged vehicle data. In the code snippet shown in the figure, images of damaged vehicles are being loaded into an in-memory CAS Table. The resulting table named "inputTable" in the example contains all the information about the images in a binary large object (blob) format.

Figure 11.13: SAS Edge Detection Applied to Damaged Vehicle Data

Load images and resize

```
conn.image.loadImages(casout={'name':'inputTable', 'replace':True}, path= path_source_images)
conn.image.processimages(casout={'name':'inputTable_resized', 'replace':True},
                    imagefunctions=[{'functionoptions':{'width':1000,'functiontype':'RESIZE','height':6
00}}],
                    imagetable={'name':'inputTable'})
imageTable = conn.CASTable('inputTable_resized')
imageShow(conn, imageTable, 0, 4)
```

NOTE: Loaded 4 images from /viyafiles/turtui/image_processing/car_accident into Cloud Analytic Services ta
ble inputTable.
NOTE: Table INPUTTABLE contains compressed images.
NOTE: 4 out of 4 images were processed successfully and saved as compressed images to the Cloud Analytic S
ervices table inputTable_resized.

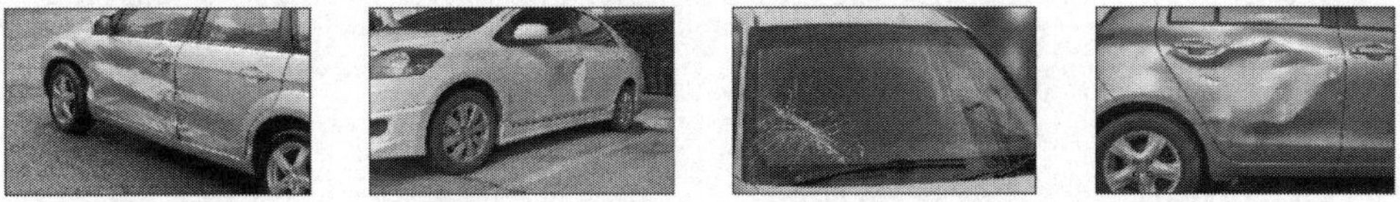

Pre-processing techniques such as resizing of the image data (shown in the example) can be applied directly via CAS actions surfaced as Python methods. Since CAS can be used on a local desktop or in a hosted cloud environment, you can analyze extremely large data sets using as much processing power as you need, while still retaining the ease-of-use of Python on the client side. Typical use cases in the insurance industry are listed below:

- Images of damaged vehicles that aid the insurer to predict the target event. Is it a write-off or a claim?
- Satellite images of homes and buildings that help in evaluating conditions of roofs for insurers at the time of underwriting. This could possibly automate scheduled inspections.

Retail

Customer Experience

Providing a personalized and enhanced customer experience in the store in real time is key in ensuring customer loyalty. Applications of machine learning, computer vision (CV), and streaming analytics covers the entire gamut from enhancing real-time, in-store customer experience, to grid surveillance and security, to improved manufacturing quality.

Facial recognition can be used to identify loyalty members as they step in the store. A camera that can scan shoppers as they enter the store combined with sophisticated deep learning models are able to identify them along with other metrics such as their gender, age, and emotion. This information could further be combined with the individual's taste profile and intelligent product recommendations could be made in real time. Facial recognition techniques can also quickly and accurately count the number of people in the store providing the data to determine the busiest days and times for a certain business. Figure 11.14 shows an implementation methodology for understanding key metrics about customers in real time.

Figure 11.14: SAS Pipeline Integrating Video Feeds from a Camera with Real-time Analytics

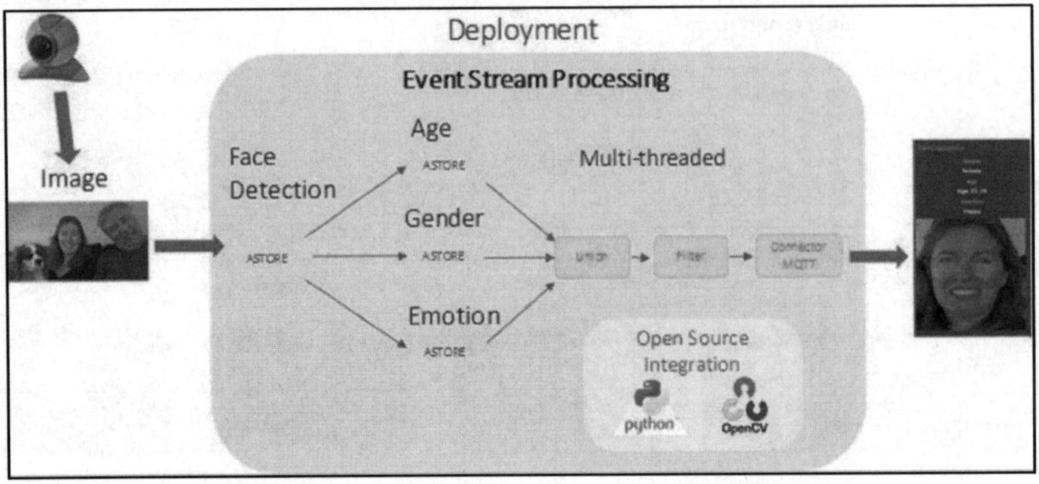

Here, a video stream is first captured by a camera. This stream is directly fed into SAS ESP Engine using the UVC connector. ASTORE files created from offline deep learning models that can identify customer's age, gender, and emotions are used to score the incoming video stream. Score results are then aggregated, enhanced, and published to an MQ Telemetry Transport (MQTT) connector. MQTT is a lightweight, publish-subscribe network protocol that transports messages between devices. To replicate the steps of this implementation framework please see https://communities.sas.com/t5/SAS-Communities-Library/Deploy-analytics-using-computer-vision-model-training-and/ta-p/582753) .

The analysis helps in making informed decisions around staffing in real time, how the customers move through the store, where customers spend the most time, and how efficient the check-out lines are. CV-enabled inventory management applications also exist that can scan products by their logo, shape, color, and other features. Finally, augmented reality applications are also being used to make recommendations about what looks good based on a customer's body type.

Data for Good

Wildlife Tracking

Applications of computer vision around Data for Good are many. Here we will cover one particular use case around wildlife monitoring. SAS works with nonprofit organizations such as WildTrack (wildtrack.org) that use non-invasive techniques to monitor endangered species.

Using digital footprints of animals, different species of animals can be identified with 90% accuracy. So far, these algorithms from SAS have been used to identify up to 15 different endangered species. This data has been collected to determine the distribution of species in their respective habitats so that conservation programs can be put in place and better strategies about which species needs monitoring can be implemented. SAS software has been used to import available images of labeled footprints and then perform a footprint classification task using SAS DLPy.

After the relevant features are learned, the deep learning model can be applied against new footprints to classify them accurately. However, accuracy is strongly dependent on the available data at hand. For detailed information, please check https://www.sas.com/en_us/explore/analytics-in-action/impact/wildtrack.html.

Figure 11.15 shows an example image of a footprint that can be ingested into the SAS deep learning framework to learn features and apply on new data.

Figure 11.15: A Footprint That Can Be Ingested by a SAS Deep Learning Model

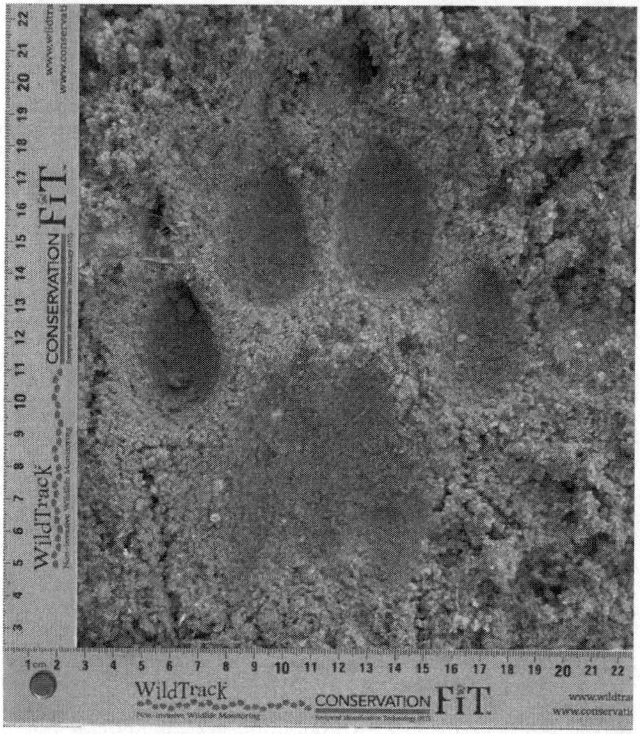

(Source: www.wildtrack.org)

Safety Compliance

Personal Protective Equipment Verification

For a multitude of industries like construction, transportation, manufacturing, or any field where worker safety is crucial, we can build object detection models to detect people and to detect whether they are wearing the requisite safety gear (vests, hard hats, shoes, and so on). We can also build models to make sure that people are not in or moving through dangerous zones. These models can be used to ensure higher safety compliance to prevent workplace injury or death.

Conclusion

Currently, there are excellent opportunities in so many different industries to build CV models with SAS Viya, deploy models for real-time scoring with SAS ESP, and integrate with existing systems to provide substantial business value. Practical use cases of AI should also improve the lives of workers and consumers and should be thought of as a tool to assist and augment human activities. As mentioned in this chapter, the use cases of computer vision in IoT span multiple industries. We have seen how SAS software plays a crucial role in every step of the IoT Analytics life cycle – data management and processing of image data, building deep learning models from scratch as well as taking advantage of transfer learning, and deployment of these models in real-time. Integration of SAS software with open-source software provides yet another level of flexibility, enabling the user to work with the best of both worlds. Armed with these tools, an AI practitioner can start to tackle any business problem in a systematic and robust manner.

References

GE Healthcare. 2018. "Beyond Imaging: the paradox of AI and medical imaging innovation." *GEHealthcare.com*. https://www.gehealthcare.com/article/beyond-imagingthe-paradox-of-ai-and-medical-imaging-innovation

Wells, J. 2018. "*5 ways Duke Energy is using drone technology*," Duke Energy Corporation. https://illumination.duke-energy.com/articles/5-ways-duke-energy-is-using-drone-technology.

Vadakkumpadan, F. and S. Sethi. 2018. "Biomedical Image Analytics using SAS Viya," *SAS Institute,* Paper SAS1961-2018.

About the Contributors

Juthika Khargharia is a Principal Solutions Architect in the SAS IoT Division specializing in machine learning, deep learning and artificial intelligence. In her role, she assists customers in defining their business problems and uses SAS advanced analytics solutions to help them reach their business goals and objectives. Juthika holds a PhD in Astrophysics and Planetary Sciences from the University of Colorado.

Hamza Ghadyali, Computer Vision Lead at the AI Center of Excellence at SAS, assists the world's largest organizations in identifying the best opportunities for leveraging AI. By building models that strategically align business goals with cutting-edge analytical capabilities, Hamza generates results that deliver value. Hamza earned his PhD in Mathematics from Duke University where he invented geometric techniques to analyze data from dynamical systems and discovered unique signatures of neuroelectrical activity during epileptic seizures.

Summary

IoT Partner Ecosystems

Previous chapters explained how SAS Event Stream Processing works and described how you can use it to apply intelligence at the edge in a variety of IoT implementation projects. In closing, it is important to note that these projects almost always require system integration, combining a range of software and hardware components. Thus, to offer complete IoT solutions, SAS partners with a variety of vendors. SAS works with these vendors in what IoT professionals call "an IoT partner ecosystem."

The following figure depicts an IoT partner ecosystem. The hardware framework includes the sensors deployed on assets to collect data and the required networking infrastructure. Within this framework, SAS software components and software from other vendors work together to complete the solution.

Figure S.1: IoT Partner Ecosystem

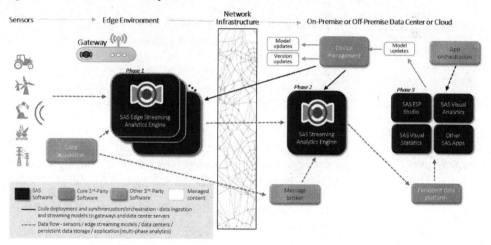

There is not a precise definition for the term *edge environment*. The figure shows aggregation points such as IoT gateways as representative edge hardware devices. It is possible for instances of fog computing to belong to the edge environment. Fog computing or fog networking drives a lot of compute power to edge devices in addition to their typical data collection activities. Thus, the architecture in the figure applies to fog computing scenarios.

The SAS software components depicted in the figure include:

- **SAS Visual Analytics and SAS Visual Statistics** – enables the exploration of large amounts of static data.
- **SAS Event Stream Processing (ESP)** – enables the real-time ingestion and analysis of the data following a streaming model definition within a streaming project. It is available for edge devices with a smaller footprint that will depend on the analytical resources needed and on data center or cloud environments integrated with SAS Viya.
- **SAS ESP Studio** – browser-based ESP streaming model development graphical interface.
- **SAS Event Stream Manager (ES Manager)** – browser-based interface that allows the management of streaming models deployed throughout the network.
- **SAS ESP Streamviewer** – browser-based interface that allows the creation of dashboards showing the data being streamed through the streaming model blocks called windows.

Two categories of software are required from other vendors: core software and optional software.

Core Software provides functionality that must be available for an end-to-end IoT solution.

Device management	Handles IoT device deployment in the areas of the following: provisioning and authentication, configuration and control, monitoring and diagnostics, and software updates and maintenance. This is critical for successful analytical deployments that leverage edge and fog computing.
Data acquisition	Provides the ability to acquire data from a variety of IoT sources that can publish data across a variety of propriety protocols. In an industrial scenario, these could be systems such as MES and Scada systems or sensors and other sampling signals that measure real-world physical conditions. This capability is critical to access data from industrial IoT devices to feed the SAS IoT analytics platform.
Persistent data platform	Includes data structures and support software that store and preserve previous versions of the streaming data incorporating additions and updates when it is modified. Existing organizational data is often added to this data lake to add context and provide a richer comprehensive source of data for discovery by SAS advanced analytics tools and solutions.

Optional Software provides additional functionality needed for specific scenarios such as multiple edge devices that interact within the same environment.

Message broker	Supports handling of combined message traffic from IoT devices. Provides reliable data orchestration services such as routing, buffering, and so on. Typically positioned as a front-end service on the data center side.
Application orchestration	Supports development of applications that are targeted toward user personals to effectively surface the insights generated by SAS IoT Analytics platform.

The activity depicted in Figure S.1 involves both multi-phased advanced data analytics and real-time streaming analytics.

- Phase 1 – streaming analytics at the edge with local context
- Phase 2 – streaming analytics at the data center with global context
- Phase 3 – advanced "data at rest" analytics at the data center ("on refresh" analytics)

These phases are sequenced in terms of where the data is first processed. Streaming analytics, especially at the edge, depends on the use case tolerance for latency and the available compute and bandwidth constraints of the network.

For Phase 1, SAS works with vendors that embed sensors into the equipment that they build. For example, consider a smart building application. Various equipment vendors for chillers, air handlers, boilers, and so on (for example, Johnson Controls or York Systems) implant sensors within their machines. Building management applications such as the Trane Tracer Summit Building Automation System coordinate the operation of this equipment and perform control functions for the building. SAS Event Stream Processing provides a connector to the BACnet protocol, which is used by these building management components and building management systems. A customer can set up alerts in a SAS Event Stream Processing model as it processes events from the building. SAS can also work with equipment and building management system vendors to customize solutions.

For Phase 2, the analytics performed at the edge are aggregated and/or filtered and then evaluated in real time at the data center. The same applications that perform control functions at the edge can be extended to a more global context.

Phase 3 places data in a persistent data platform (Hadoop for example) and analyzes it based on the current content. The timing of data placement depends on the update strategies and methods of the platform (real time, near real time, and so on).

As SAS works with vendors, it builds partnerships with them. For example, Trane provides an analytics system running in its cloud. SAS partners with Trane to provide advanced analytic routines within that cloud. SAS also works with IoT platform vendors such as GE Predix, Siemens, or PTC ThingWorx that can cover end-to-end functionality. GE Predix and Siemens have partnered with SAS to embed its analytics components within their solutions. Siemens Mindsphere, which covers typical industrial use cases including predictive asset maintenance, now embeds SAS streaming analytics to meet growing demands for IoT analytics with AI and machine learning capabilities

An IoT partner ecosystem also includes system integrators such as Wipro and Accenture. Wipro generally assigns personnel to projects, whereas Accenture generally focuses on higher level system design. Both assemble an IoT implementation from multiple vendors and supervises its implementation.

With this added perspective, it should be clear how SAS products in general, and SAS Event Stream Processing in particular, provide a keystone to complete IoT solutions. The previous chapters were written to explicate the business value that you can obtain from the application of intelligence to the edge through SAS Event Stream Processing. As the number of "things" grows, SAS will find new and better ways to enable users to harness the business value of that applied intelligence. And as you have discovered by reading this book, the edge is everywhere.

Additional Resources

You can refer to the following sources for more information about SAS Event Stream Processing, related SAS products, and the Internet of Things.

The Learn and Support page for SAS Event Stream Processing (https://support.sas.com/en/software/event-stream-processing-support.html) provides access to product documentation, tutorials, information about training courses, links to relevant blogs and online communities, software hot fixes, and other support material.

The ESPPy package, available at https://github.com/sassoftware/python-esppy, enables you to create SAS Event Stream Processing models programmatically in Python. For information about how to get started with ESPPy, refer to https://communities.sas.com/t5/SAS-Communities-Library/ESPPy-How-to-get-started-with-the-Python-package-for-SAS-Event/ta-p/513050.

You can test drive SAS Event Stream Processing in the SAS Analytics Cloud (https://analyticscloud.sas.com/). After you set up a user profile, you can use the trial programming environment to build streaming projects using real-time data. You can run ESPPy projects in Jupyter Lab, run SAS ESP Studio, or run SAS ESP Streamviewer.

The SAS Deep Learning Python (DLPy) package, available at https://github.com/sassoftware/python-dlpy, provides the high-level Python APIs to deep learning methods in SAS Visual Data Mining and Machine Learning. For help with getting started with DLPy, refer to (https://video.sas.com/detail/video/6001746946001/deep-learning-with-python-dlpy-and-sas-viya---introduction-to-the-series).

For more information about how to build computer vision models with Python and DPLy, refer to https://blogs.sas.com/content/subconsciousmusings/2019/09/24/videos-computer-vision-models/.

For more information about how to use deep learning with SAS, including steps for building a deep learning model, refer to https://www.sas.com/content/dam/SAS/en_us/doc/whitepaper1/deep-learning-with-sas-109610.pdf.

Session proceedings from SAS Global Forum are available at https://www.sas.com/en_us/events/sas-global-forum/program/proceedings.html. Some of those papers served as the basis of chapters for this book.

Other papers of interest include the following:

- "SWAT's it all about? SAS Viya® for Python Users" by Carrie Foreman, Amadeus Software Limited. Paper 3610-2019. https://www.sas.com/content/dam/SAS/support/en/sas-global-forum-proceedings/2019/3610-2019.pdf.
- "Medical Image Analytics in SAS® Viya® with Applications in the Treatment of Colorectal Cancer Spread to the Liver" by Fijoy Vadakkumpadan and Joost Huiskens, SAS Institute Inc. Paper SAS3341-2019. https://www.sas.com/content/dam/SAS/support/en/sas-global-forum-proceedings/2019/3341-2019.pdf.
- "Exploring Computer Vision in Deep Learning: Object Detection and Semantic Segmentation" by Xindian Long, Maggie Du, and Xiangqian Hu, SAS Institute Inc. Paper 3317-2019. https://www.sas.com/content/dam/SAS/support/en/sas-global-forum-proceedings/2019/3317-2019%20.pdf.
- "Real-Time Image Processing and Analytics Using SAS® Event Stream Processing" by Frederic Combaneyre, SAS Institute Inc. Paper SAS2103-2018. https://www.sas.com/content/dam/SAS/support/en/sas-global-forum-proceedings/2018/2103-2018.pdf
- "Creating a Strong Business Case for SAS® Event Stream Processing" by Henrique C. Danc, SAS Institute Brazil, Sao Paulo, Brazil. Paper SAS3201-2019. https://www.sas.com/content/dam/SAS/support/en/sas-global-forum-proceedings/2019/3201-2019.pdf.
- "Augmented Reality (AR) Predictive Maintenance System with Artificial Intelligence (AI) for Industrial Mobile Robot" by Yee Yang Tay, Kai Woon Goh, Marvin Dares, Ye Sheng Koh, and Che Fai Yeong, Universiti Teknologi Malaysia; Mark Chia, SAS Institute; Ping Hua Tan, DF Automation and Robotics Sdn Bhd. Paper 3628-2019. https://www.sas.com/content/dam/SAS/support/en/sas-global-forum-proceedings/2019/3628-2019.pdf.

Ready to take your SAS® and JMP®skills up a notch?

Be among the first to know about new books,
special events, and exclusive discounts.
support.sas.com/newbooks

Share your expertise. Write a book with SAS.
support.sas.com/publish